Calculus I

Resource for Peer-Led Team Learning Workshop

Editors

June L. Gastón, Ed.D., Professor Emerita
City University of New York

Franklin G. Saladini, Peer Leader
University of Texas Rio Grande Valley

Karmen T. Yu, Ph.D.
Montclair State University

Contributors

Peggy Beck, Prince George's Community College
Abdelkrim Brania, Morehouse College
Paula Drewniany, University of Maine, Orono
John Merkel, Morehouse College
Jennifer Tyne, University of Maine, Orono

Published by the Peer-Led Team Learning International Society

A non-profit organization for the advancement of learning

Calculus I: Resource for Peer-Led Team Learning Workshop

Copyright ©2025 by Peer-Led Team Learning International Society

All rights reserved.

Permission in writing must be obtained from Peer-Led Team Learning International Society before any part of this work is reproduced or copied in any form or by any means, electronic or mechanical, including photographing, photocopying and recording, or by any information storage or retrieval system.

Cover Art: Photograph by AE Dreyfuss
Little Island, Hudson River Park, New York City, New York

Cover Design: Jonathan G. Lopez

10 9 8 7 6 5 4 3 2 1

ISBN 978-1-944996-10-9

Peer-Led Team Learning International Society (www.pltlis.org)

7100 Westwind Drive, Suite 240, El Paso, Texas 79912 USA
info@pltlis.org

Series Editors: A.E. Dreyfuss, Ed.D.
James E. Becvar, Ph.D.
Ana Fraiman, Ph.D.
Karmen T. Yu, Ph.D.

Table of Contents

Acknowledgments	1
Preface	3
Message To the Student	4
Topic 1: Introduction to Calculus I	7
Exploratory Workshop: Calculus Themes	7
Topic 2: Limits	17
Exploratory Workshop: Calculating Limits	17
Topic 3: Continuity	21
Exploratory Workshop: Rethinking Continuity	21
Topic 4: Introduction to the Derivative	25
Discovery Workshop: Rates of Change and Tangents	25
Topic 5: The Derivative	31
Review Workshop: A Conceptual Review of the Derivative	31
Topic 6: Velocity	39
Discovery Workshop: Velocity/Rates	39
Topic 7: Differentiation	51
Review Workshop: Differentiation Rules	51
Topic 8: The Chain Rule	57
Workshop: Composition of Functions and the Chain Rule	57
Topic 9: Higher Order Derivatives and Implicit Differentiation	63
Review Workshop: A Comprehensive Look at Higher Order Derivatives and Implicit Differentiation	63
Topic 10: Related Rates	69
Review Workshop: Related Rates with Geometry and Modeling	69
Topic 11: Optimization	77
Review Workshop: Optimization with Geometry and Modeling	77
Topic 12: Surge Functions	85
Exploratory Workshop: Modeling with Surge Functions	85
Topic 13: Antiderivatives	93
Exploratory Workshop: Reversing Differentiation	93
Topic 14: Indefinite Integrals	101
Review Workshop: Meeting Challenges of the Indefinite Integral	101
Topic 15: Riemann Sums	105
Review Workshop: Riemann Sums	105
Topic 16: The Fundamental Theorem of Calculus	115
Review Workshop: The FTC	115

Acknowledgments

These PLTL workshop materials aim to challenge students so that discussion among group members helps them arrive at solutions through collaborative processes. Solving problems in pairs, trios, or quartets, and using facilitation strategies can generate great conversations over a simple mathematical problem. When students are engaged with the problems and discussing the concepts, they are discovering the many ways of learning.

This resource book is based on materials originally developed by a diverse group of authors: Peggy Beck of Prince George's Community College, Paula Drewniany and Jennifer Tyne of the University of Maine, Orono, and Abdelkrim Brania and John Merkel of Morehouse College.

Robert J. Alvarez, former Peer Leader at the University of Texas El Paso, was instrumental in editing an earlier draft of this workbook.

It is anticipated that the use of these materials will continue to both enable and support Peer-Led Team Learning and students' learning.

- The Series Editors

Preface

This workbook is designed to provide structure to the Peer-Led Team Learning (PLTL) Workshops for the college Calculus I course. PLTL mathematics workshops are based on the Peer-Led Team Learning (PLTL) program. PLTL is a proven method for improving learning for college students. A small group of students is led by a qualified and trained student, a peer leader.

Originally developed from courses at the University of Maine, Morehouse College, and Prince George's Community College in Maryland, these materials have been tested and provide students with many valuable tools for success.

Important tools for students include:

- How to approach and improve learning
- How to explore concepts of college algebra
- How to ask the right kind of questions
- How to cooperate with each other and learn from peers
- How to develop mathematical understanding that prepares them for their current and future coursework
- How to study effectively and efficiently
- How to develop a network of reliable colleagues that can enrich their learning environment.

Learning these skills is essential for students to succeed in higher education.

The use of this resource book within the workshop environment facilitates the implementation of workshop activities and reinforces essential learning modes. The peer leader can apply the material included in this workbook for use during the workshop throughout the semester. If your course does not include a Peer-Led workshop, you will find these additional materials helpful to develop critical thinking in solving Calculus problems. You may also benefit by creating a study group with classmates and using this workbook to guide your learning.

Message To the Student

Welcome to the Peer-Led Team Learning workbook for Calculus I. This resource book presents various kinds of problems, which are presented in one of three categories. These may be:

(1) a **discovery workshop**, best offered before discussion of a concept occurs in class, with the goal that your learning will be enhanced through discovering an idea with peers;

(2) an **exploratory workshop**, best offered after a concept is discussed in class so you can explore concepts in more depth with the goal of deepening your understanding of the idea;

(3) a **review workshop** that is offered so you can practice important skills. You may be asked to complete the entire workshop, or selected parts because of the extensive nature of the topic and/or time constraints.

Strategies for Success

Calculus was developed to address real life mathematical problems involving motion. Success in the study of Calculus can support conceptual and procedural understanding of subsequent mathematical coursework in areas such as the behavioral and social sciences, biology and medicine, business and economics. It becomes challenging if you do not understand key concepts and do not ask for help, or do not get the right help. The most important action you can do to ensure your success in the course is to actively participate in each lecture and to avail yourself of opportunities to maximize understanding, by participating in peer-led workshops or study groups, consulting your professor during office hours, and utilizing tutoring options.

In the study of mathematics, it is important to develop problem-solving skills that facilitate analytical and critical thinking. Collaborating with other students and getting help from mentors are proven strategies to improve these problem-solving skills. Please take advantage of the resources that are provided to you from your institution. In addition, keep your eyes and mind open to new learning experiences that a college education offers.

Networking Facilitation Page: Your Workshop Group

Professor: _____   Office Hours: _____

Professor's Email: _____   Professor's Office: _____

Peer Leader: _____   Peer Leader's Office: _____

Peer Leader's Email: _____   Phone Number: _____

Workshop Code: _____ Day: _____ Time: _____

Contacts

Name	Email

My Notes

Topic 1: Introduction to Calculus I

Exploratory Workshop: Calculus Themes

How big is the largest rubber band ball? Former peer leader John Bain created this gigantic rubber band ball from 850,000 rubber bands. It is approximately 5 feet in diameter and weighs over 3,000 pounds. He started with handfuls of elastics that were free from the post office, and the ball kept growing! He also inspired others who later broke his Guinness record for the largest ball.

PART 1.

1. Inspired by John Bain's story, you decide to construct your own rubber band ball. Every day you add 5 rubber bands to the ball. What is the rate of change (in rubber bands per day) of the number of rubber bands on the ball?

2. If you start with zero rubber bands, complete the following table for days 1 – 10:

Day	Rate at which the Rubber Band Ball is Growing (rubber bands/ day)	Number of Rubber Bands in the Ball
1		
2		
3		
4		
5		
6		
7		
8		
9		
10		

3. On the following graph, plot the data points (day, number of rubber bands) from the previous table.

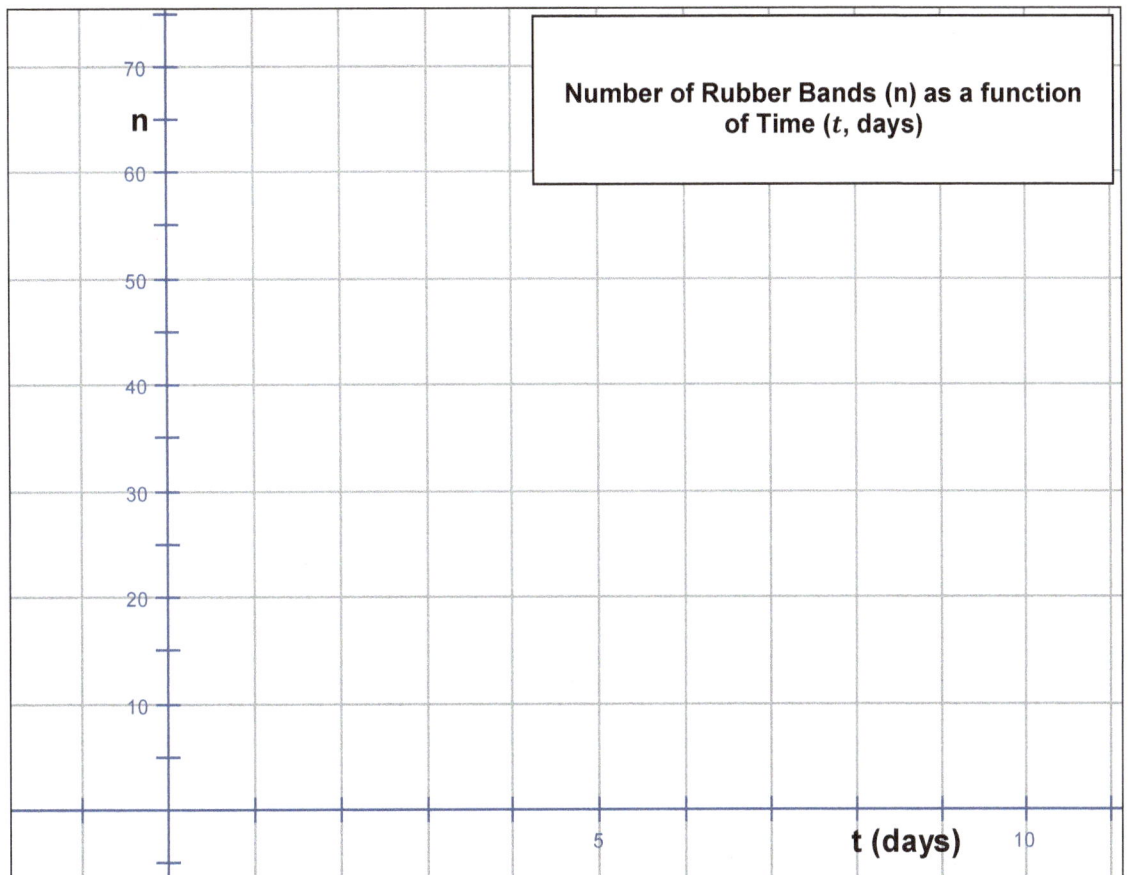

Number of Rubber Bands (n) as a function of Time (t, days)

4. A smooth curve drawn through these data points would have what shape?

5. What is the slope of the curve? What does it represent?

You decide to abandon your rubber band ball effort and begin building a paper clip chain. You start with 10 paper clips on your chain. In the first week you add 3 paper clips to the chain. During the second week you add 6 paper clips to the chain. During the third week you add 9 paper clips to the chain. Each week the number of paper clips you add to the chain is three more than the number you added the week before.

6. Complete the following table for weeks 0 – 10:

Week	Rate at which the Paper Clip Chain is Growing (paper clips/ week)	Number of Paper Clips in the Chain
0	-	10
1	3	
2	6	
3	9	
4		
5		
6		
7		
8		
9		
10		

7. What is the rate of change (in paper clips per week) of the number of paper clips on the chain? What makes this question more difficult to answer than question 1?

8. If you could not answer the general question in question 7, how about this one: In the tenth week what is the rate of change (in paper clips per week) of the number of paper clips on the chain?

9. Plot the data points (week, number of paperclips) from the table in question 6 on the graph below, and draw a smoothed curve through the data points

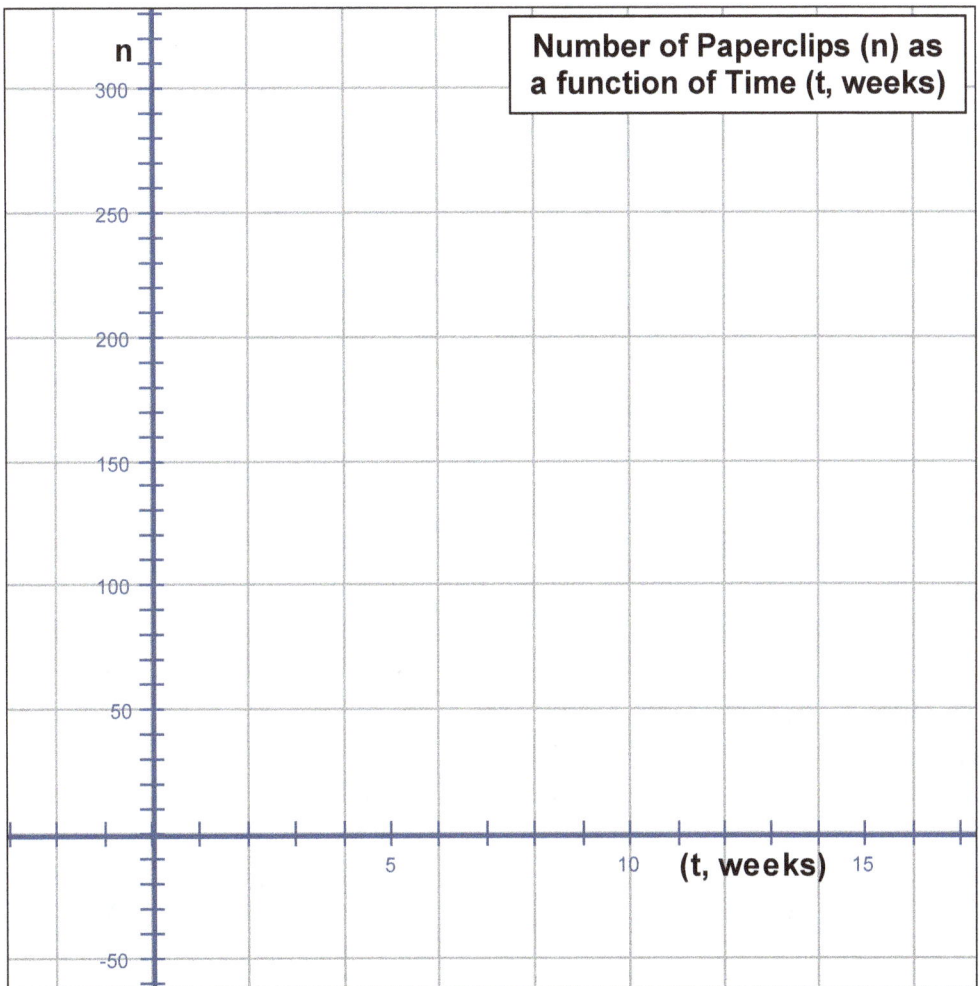

Number of Paperclips (n) as a function of Time (t, weeks)

10. How does this graph differ from the one for the rubber bands?

11. How is this difference related to the rates of change of the two quantities?

12. How would you describe the slope of the curve?

13. The graph of the equation $P = 2.5t + 10$ is shown on the right.

What is the rate of change of P with respect to t?

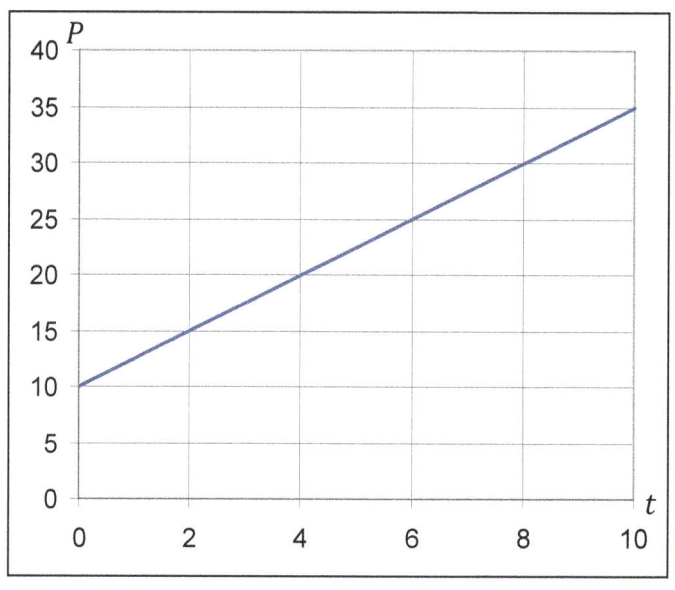

14. The graph of the equation $y = -x^2 + 9x + 6$ is shown on the right.

Consider the question: What is the rate of change of y with respect to x?

Can you answer it? If yes, what is the answer; if no, what causes difficulties in trying to answer it?

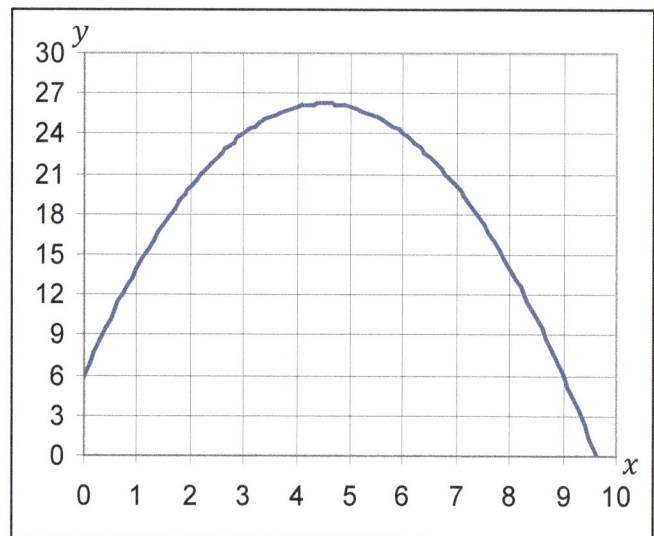

15. Instead consider the question: What is the rate of change of y with respect to x when $x = 2$? Discuss how you could answer that using the graph provided above. Try to answer it if you can.

PART 2.

1. A farmer has a field as shown on the right. The measurements are in feet. What is the area of the farmer's field?

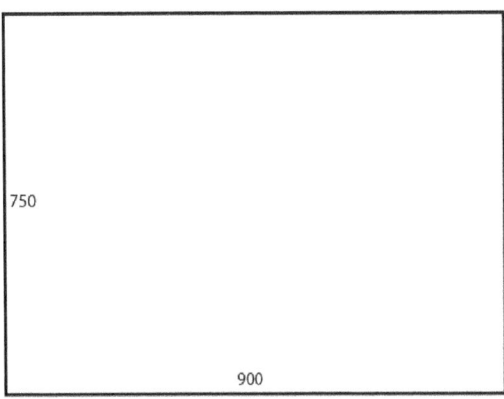

2. A farmer has a field as shown on the right. One side is bordered by a river, and the side opposite the river is bordered by a road. The other two bordering sides of the field run perpendicular to the road, from the road to the river. The farmer wishes to approximate the area of her field, so at 100-foot intervals along the road, she measured the distance (in feet) from the road to the river. Her measurements are shown on the right.

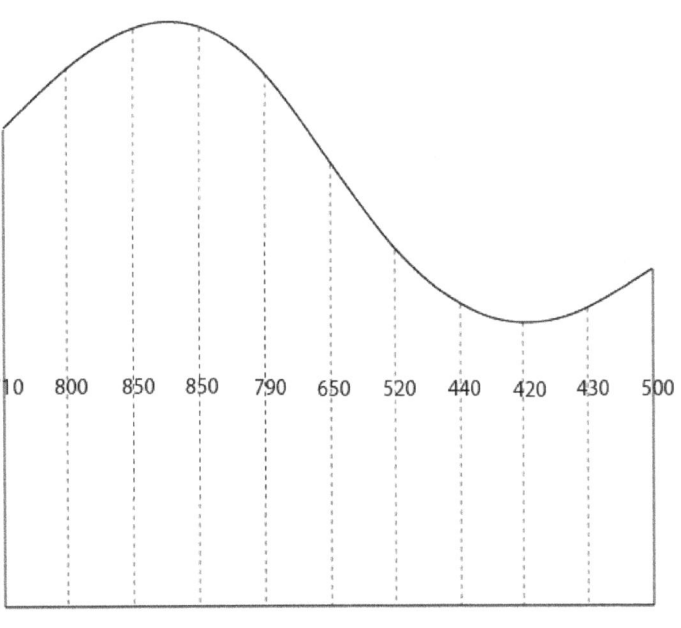

Use the information provided to approximate the area of the field and explain how you arrived at your approximation.

Discuss what could be done to make a more accurate approximation of the area.

13

3. The graph of the equation $y = 20$ is shown on the right.

   Shade in the area between $x = 2$ and $x = 9$ that is bounded above by the graph and bounded below by the x axis.

   What is the area of the figure you shaded?

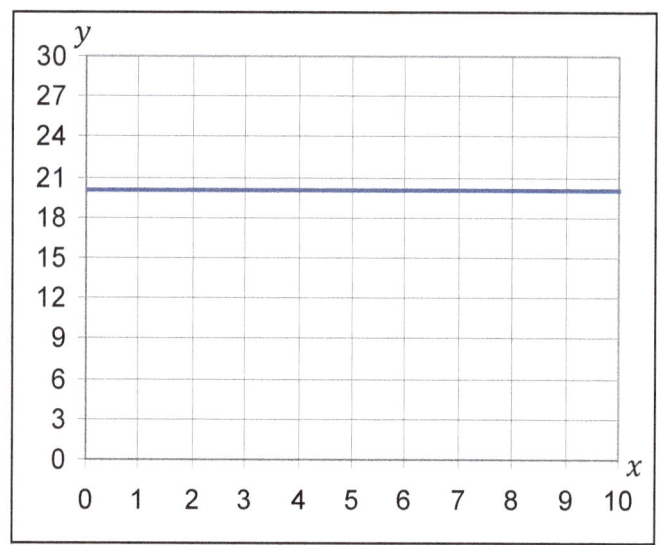

4. The graph of the equation $y = -x^2 + 9x + 6$ is shown on the right.

   Shade in the area between $x = 2$ and $x = 9$ that is bounded above by the graph and bounded below by the x axis.

   Estimate the area of the figure you shaded and discuss how you made your estimate.

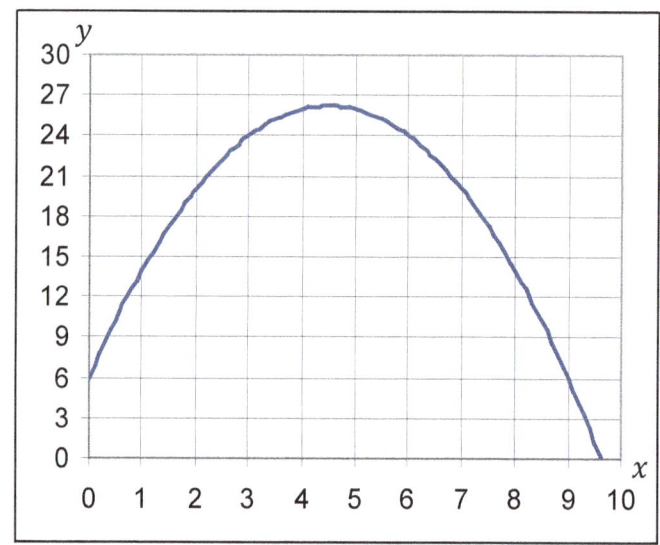

Discuss how you could improve your estimate.

PART 3.

Conclusion

The goal of this workshop was to investigate the two central themes of Calculus. Briefly summarize what those ideas are.

Identify the problems in this workshop that most helped you understand those ideas.

My Notes

Topic 2: Limits

Exploratory Workshop: Calculating Limits

The concept of a limit is fundamental to the development of calculus and is in fact what distinguishes calculus from other branches of mathematics. We can analyze limits from a numerical approach (make a table of values), a graphical approach (trace the graph to see what value *y* nears), and an algebraic approach (simplify and substitute). What other approaches can we add to this list?

Part 1.

For each of the following problems, (1) decide which method(s) can be used, and (2) calculate the limit by each applicable method.

1. $\lim\limits_{x \to 4} \dfrac{\sqrt{x^2 - 8x + 16}}{x - 4}$

method	numerical	graphical	algebraic
limit computation			

2. $\lim\limits_{x \to 0} \dfrac{\sin(x)}{x}$

method	numerical	graphical	algebraic
limit computation			

Part 2.

Sometimes other methods are needed. The "squeeze theorem" says that if you can bound your function both above and below by well-behaved functions that both converge to the same limit, then your function converges to that same limit. This can be formally stated:

If $g(x) \leq f(x) \leq h(x)$ for all $x \neq a$ in some interval about a, and $\lim_{x \to a} g(x) = \lim_{x \to a} h(x) = L$, then $\lim_{x \to a} f(x) = L$.

1. Let's look at $\lim_{x \to 0} x^2 \cos\left(\frac{1}{x^2}\right)$. Compare it to the functions $y = x^2$ and $y = -x^2$. In terms of the Theorem above, which function is $g(x)$ (the lower bound) and which is $h(x)$ (the upper bound)?

   Find $\lim_{x \to 0} x^2$ and $\lim_{x \to 0} -x^2$. What conclusions can you draw?

2. Explore the limit $\lim_{x \to 0^-} \frac{[\![x]\!]+1}{x}$, where $[\![x]\!]$ is the greatest integer function (y is the greatest integer less than or equal to x). Consider whether the functions $g(x) = \frac{x}{x-1}$ and $h(x) = \frac{x}{x+1}$ work as upper and lower bounds.

   Indicate which one is an upper bound and which one is a lower bound.

   Indicate which of these functions converges to 0 from the left.

Part 3.

1. In Calculus II, you will study "series", which are sums. For example,
$1 + \frac{1}{2^2} + \frac{1}{3^2} + \frac{1}{4^2} + \frac{1}{5^2} + \ldots + \frac{1}{n^2} + \ldots$ is written as $\sum_{n=1}^{\infty} \frac{1}{n^2}$. How can you use your calculator to compute these sums?

Hint: The formula is $\text{sum}\left(\text{seq}\left(\frac{1}{(x^2)}, x, 1, \#\right)\right)$, where # is the number of terms you wish to add.

Use this formula to compute the sum $\sum_{n=1}^{3} \frac{1}{n^2} = 1 + \frac{1}{2^2} + \frac{1}{3^2}$ (in the formula, use # = 3) and check by hand.

Compute the sum for each of the following where # indicates the number of terms:

= 10 $\sum_{n=1}^{10} \frac{1}{n^2} =$

= 100 $\sum_{n=1}^{100} \frac{1}{n^2} =$

= 500 $\sum_{n=1}^{500} \frac{1}{n^2} =$

= 800 $\sum_{n=1}^{800} \frac{1}{n^2} =$

= 900 $\sum_{n=1}^{900} \frac{1}{n^2} =$

Based on this information, try to guess $\lim_{k \to \infty} \sum_{n=1}^{k} \frac{1}{n^2}$. As you add more terms, is there a particular value that the sum seems to approach?

2. Here is another series:
$1 + \frac{1}{2} + \frac{1}{3} + \frac{1}{4} + \frac{1}{5} + \ldots + \frac{1}{n} + \ldots$ is written as $\sum_{n=1}^{\infty} \frac{1}{n}$. How can you use your calculator to compute these sums?

Hint: The formula is $\text{sum}\left(\text{seq}\left(\frac{1}{x}, x, 1, \#\right)\right)$, where # is the number of terms you wish to add.

Use this formula to compute the sum $\sum_{n=1}^{3} \frac{1}{n} = 1 + \frac{1}{2} + \frac{1}{3}$ (in the formula, use # = 3) and check by hand.

Compute the sum for each of the following where # indicates the number of terms:

$\# = 10$ $\qquad$ $\sum_{n=1}^{10} \frac{1}{n} =$

$\# = 100$ $\qquad$ $\sum_{n=1}^{100} \frac{1}{n} =$

$\# = 500$ $\qquad$ $\sum_{n=1}^{500} \frac{1}{n} =$

$\# = 800$ $\qquad$ $\sum_{n=1}^{800} \frac{1}{n} =$

$\# = 900$ $\qquad$ $\sum_{n=1}^{900} \frac{1}{n} =$

Based on this information, try to guess $\lim_{k \to \infty} \sum_{n=1}^{k} \frac{1}{n}$. As you add more terms, is there a particular value that the sum seems to approach?

Topic 3: Continuity

Exploratory Workshop: Rethinking Continuity

The notions of limit and continuity are intimately related. In fact, continuity is defined in terms of a limit, and if a function f is continuous at a point c then it must have a limit at that point. In general, limits and continuity are important concepts to use in the analysis of a function.

Part 1.

A function is **continuous** at the point $x = a$ if three conditions are met:

    i. $\lim_{x \to a} f(x)$ exists

    ii. $f(a)$ is defined.

    iii. $\lim_{x \to a} f(x) = f(a)$

1. When is a function continuous over an interval?

2. Is the function $f(x) = \frac{\sin x}{x}$ continuous at $x = 0$? Why/why not?

3. Is the function $f(x) = [\![x]\!]$ (the greatest integer function) continuous at $x = 1$? Why/why not?

4. Is the function $f(x) = \begin{cases} 3x^2 + 2x - 1; & x \neq 1 \\ 5; & x = 1 \end{cases}$ continuous at $x = 1$? Why/why not?

5. Is the function $f(x) = \begin{cases} 3x^2 + 2x - 1; & x \neq 1 \\ 5; & x = 1 \end{cases}$ continuous at $x = 2$? Why/why not?

6. Is the function $f(x) = \begin{cases} x^2; & x \leq 2 \\ x^2 - 8x + 16; & x > 2 \end{cases}$ continuous at $x = 2$? Why/why not?

7. Find the constant c that will make $f(x)$ continuous at $x = 1$. $f(x) = \begin{cases} cx^2 - 1; & x < 1 \\ -2x - 4; & x \geq 1 \end{cases}$

Part 2.

There are three basic types of discontinuities (places where the function is not continuous):

  i. **Removable** - The function can be redefined at $x = a$ to make it continuous.

  ii. **Jump** - The function does not have a limit as x approaches a because the left-hand limit does not equal the right-hand limit.

  iii. **Essential** - The function does not have a limit as x approaches a because either the left-hand limit or the right-hand limit (or both) does not exist.

1. Which functions from Part 1 were discontinuous?

2. Label each function below with the correct type of discontinuity.

   a) $f(x) = \frac{\sin x}{x}$ at $x = 0$

   b) $f(x) = [\![x]\!]$ at $x = 1$

   c) $f(x) = \begin{cases} 3x^2 + 2x - 1; x \neq 1 \\ 5 \qquad\qquad ; x = 1 \end{cases}$ at $x = 1$

d) $f(x) = \begin{cases} 3x^2 + 2x - 1; & x \neq 1 \\ 5; & x = 1 \end{cases}$ at $x = 2$

e) $f(x) = \begin{cases} x^2; & x \leq 2 \\ x^2 - 8x + 16; & x > 2 \end{cases}$ at $x = 2$

Part 3.

The definitions for **continuity** and **limit** are sometimes confusing. Carefully read/compare the definitions in your text or notes. Write a note to the desperate student, I.M. Flummoxed, clearly explaining the two terms, including the differences between the two concepts. Indicate at least two problems in this workshop that help clarify each concept.

Topic 4: Introduction to the Derivative

Discovery Workshop: Rates of Change and Tangents

Part 1.

To better understand the concept of *derivative*, we will explore slopes and tangent lines.

Let $f(x) = x^3 - 5x^2 + 2$.

1. Graph the function.

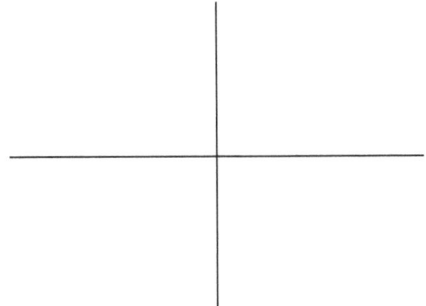

2. Find the zeroes (roots) of $f(x)$ to two decimal places. Mark the points on your graph above.

3. Find the relative minimums and maximums. Mark the points on your graph above.

4. Find $g(x) = \lim\limits_{h \to 0} \frac{f(x+h)-f(x)}{h}$. (Hint: Find $(x+h)^3 - 5(x+h)^2 + 2$, etc.)

5. The relative rate of change of $f(x)$ at points x is measured by the slope of the tangent line. Find the slopes at these x-values.

x	−1	−0.5	0	0.5	1	2	3	3.3	4.5
$g(x)$									

Note that the slopes change sign at $x = 0$ and at $x = 3.3$.

6. Graph the function $g(x) = 3x^2 - 10x$, indicating its x-intercepts.

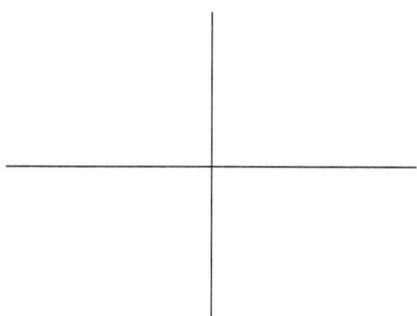

7. Observe that $g(x) = 3x^2 - 10x$ is positive on the intervals $(-\infty, 0)$ and $(3.\bar{3}, \infty)$, and is negative on the interval $(0, 3.\bar{3})$. These intervals correspond to the signs (positive and negative) of the slope of $f(x)$. What traits do $f(x) = x^3 - 5x^2 + 2$ have on these intervals?

8. What is the relationship between the zeroes of $g(x) = 3x^2 - 10x$ and the extrema (minimums and maximums) of $f(x) = x^3 - 5x^2 + 2$?

> We will be calling $g(x)$ the **derivative** of $f(x)$. We will write $g(x)$ as $f'(x)$. We will also study the connections of $g(x)$ to the maximums and minimums of $f(x)$.

9. Use the slopes of the function $f(x) = x^3 - 5x^2 + 2$ computed in #5 above to find the equations of the tangent lines to $f(x)$ at the points where:

   a) $x = -1$

   b) $x = 0$

   c) $x = 1$

10. Refer back to #9.

    a) Graph the function $f(x)$ with the three tangent lines.

    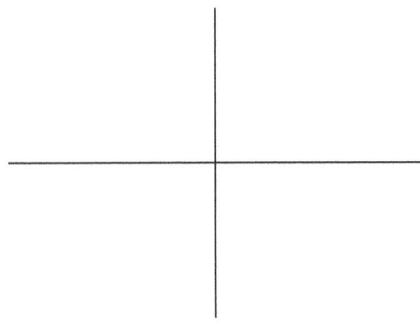

    b) In which interval is the slope positive? In which interval is the slope negative?

    c) Use the slope to describe the behavior of the graph of the function.

Part 2.

Let $f(x) = x^2(x+1)(x-1)(x-2) = x^5 - 2x^4 - x^3 + 2x^3$.
Then, $f'(x) = 5x^4 - 8x^3 - 3x^2 + 4x$.
Both $f(x)$ and $f'(x)$ are graphed below over the interval $[-1.5, 2.5]$

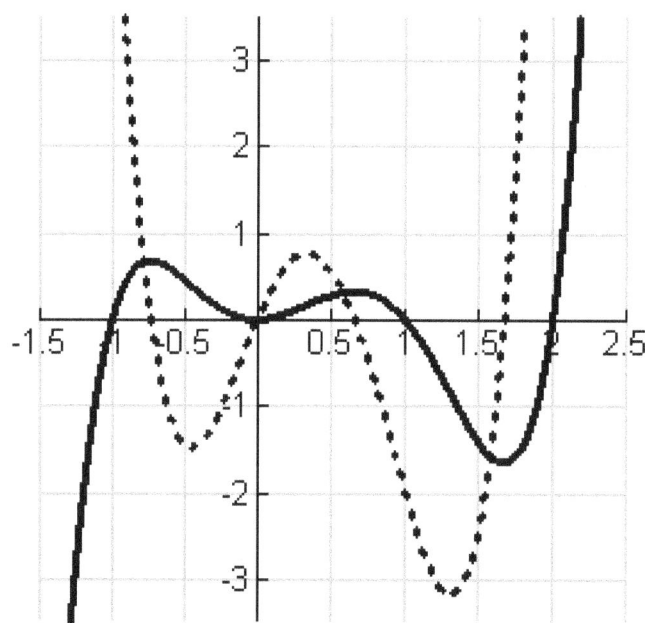

1. Which is the graph of $f(x)$, and which is the graph of $f'(x)$? Label the graph.

2. Answer the following questions by inspection of the graph:

   a) Over what intervals is $f(x)$ increasing?

   b) Over what intervals is $f'(x) > 0$?

   c) Over what intervals is $f(x)$ decreasing?

   d) Over what intervals is $f'(x) < 0$?

   e) What are the x-coordinates of all the extrema of the graph of $f(x)$?

f) For what values of x is $f'(x) = 0$?

3. Refer back to your answers to Part 2, problems 1 and 2. Write a statement that relates the behavior of a function (where it is increasing, decreasing, has maximum and minimum points) to properties you have observed about the graph of its derivative.

My Notes

Topic 5: The Derivative

Review Workshop: A Conceptual Review of the Derivative

The central concept in differential calculus is the derivative of a function. It is the limit of the average of the function over smaller and smaller intervals. The derivative of a function plays an important role in describing both analytically and geometrically the rate of variation between two or more variables. Many dynamic processes found in nature can be described in terms of derivatives. For example, heat conduction, power transmission, and population growth are all studied through equations that involve derivatives of a function.

Part 1A. Basic terminology.

2. Define each term below.

   a) Derivative at a point

   b) Difference Quotient

3. Explain your conceptual understanding. (Be sure to use appropriate mathematical notation in your explanation.)

   a) What is the connection between the **difference quotient** and the **average rate of change of the function**?

   b) (b) What is the connection between the **derivative at a point** and the **difference quotient**?

Part 1B. Using the definition of the derivative, solve each problem.

1. Let $f(x) = \sqrt{x}$. Find $f'(3) = \lim_{h \to 0} \frac{f(3+h)-f(3)}{h}$.

2. Let $f(x) = \frac{1}{x+2}$. Find $f'(0) = \lim_{h \to 0} \frac{f(0+h)-f(0)}{h}$.

3. Let $f(x) = \sin(x)$. Use the addition formula $\sin(x+h) = \sin(x)\cos(h) + \cos(x)\sin(h)$ and the two limits involving trigonometric functions to show that.

$$f'(c) = \lim_{h \to 0} \frac{\sin(x+h) - \sin(x)}{h} = \cos(x)$$

Part 2. Focus on **differentiability** versus **continuity**.

1. Explain your conceptual understanding.

    a) When is a function differentiable at a point?

    b) How does differentiability at a point x imply continuity at the point x?

c) Is $f(x) = |x|$ continuous at $x = 0$?

2. Show that $f(x) = |x|$ is not differentiable at $x = 0$, that is, show $f\,'(0)$ does not exist. (Hint: calculate both $\lim\limits_{h \to 0^-} \frac{f(0+h)-f(0)}{h}$ and $\lim\limits_{h \to 0^+} \frac{f(0+h)-f(0)}{h}$).

3. Based on your answer to problem 2, does continuity at a point imply differentiability at that point?

4. Is $g(x) = \begin{cases} x^2 & \text{if } x \leq 1 \\ 2x & \text{if } x > 1 \end{cases}$ continuous at $x = 1$? Explain why or why not.

5. Show that $\lim\limits_{h \to 0^+} \frac{g(1+h)-g(1)}{h}$ does not exist. Be careful when calculating $g(1)$.

6. Based on your answer to problem 5, is g differentiable at $x = 1$?

7. Generalizing your results in problems 5 and 6, can a function that is not continuous at a point be differentiable at that point?

Part 3. Derivative as the slope of a tangent line

1. Review terms and concepts

    a) Define the slope of the secant line in terms of the difference quotient.

    a) How can we find the slope of the graph of a function f at $x = c$?
       (Hint: Explain how the tangent line and the secant line are used.)

    b) If $f'(c)$ exists, the graph of the function f has a tangent line at $(c, f(c))$. Explain whether the converse of this statement is or is not always true.

2. Let $f(x) = x^2$.

    a) Sketch the graph of $y = f(x)$ for $-3 \leq x \leq 3$.

    b) Compute $f'(1)$.

c) Write the equation of the tangent line to the graph of f at the point $(1, f(1)) = (1, 1)$.

d) Write the equation of the tangent line as a linear function $L(x)$, and use it to approximate each value $f(x)$ in the table:

x	$f(x)$	$L(x)$	$|f(x) - L(x)|$
1.8			
1.5			
1.2			
1.1			
1.05			
1.01			
1.001			

e) By observing the error values found above, what can you conclude about approximating the graph of f near $c = 1$ by the tangent?

3. Let $f(x) = x^{1/3}$.

a) Sketch the graph of f for $-8 \leq x \leq 8$.

b) Show why $f'(0) = \lim\limits_{h \to 0} \dfrac{f(0+h) - f(0)}{h}$ does not exist.

c) Is the graph of f continuous at $x = 0$?

d) From your graph, what is the slope of the tangent line at the point $(0, f(0)) = (0,0)$.

e) Based on your results, if a function has a tangent line at a point, is it necessarily the case that the function is differentiable at that point?

Part 4. Derivative as a rate of change

1. Review formulas. Write the formula for each of the following terms.

    a) Velocity $v(t)$ of an object at the time t

    b) Derivative of the position function

    c) Acceleration $a(t)$ of an object at the time t

    d) Derivative of the velocity function

2. Which of the above formulas designate equality? Why?

3. An object is thrown vertically upward from a height of 64 feet. The position of the object at time t is given by $s(t) = -16t^2 + 48t + 64$ for $t \geq 0$. ($s(t)$ in feet and t in seconds)

   a) Find the velocity $v(t)$ of the object at any time $t \geq 0$.

   b) Find the acceleration $a(t)$ of the object at any time $t \geq 0$.

   c) What is the initial position $s(0)$ and initial velocity $v(0)$ of the object?

   d) Determine the time at which the object hits the ground.

   e) With what velocity does the object hit the ground? Explain.

4. The volume V of a sphere is related to the radius r of the sphere by the equation $V(r) = \frac{4}{3} \cdot \pi r^3$.

   a) Find the rate of change of volume of the sphere with respect to its radius; $V'(r) =$

   b) Use the formula found above to complete the table below. What do the values in your table represent?

r	$V'(r)$
1	
2	
5	
10	

Topic 6: Velocity

Discovery Workshop: Velocity/Rates

Introduction

Bob and Sue are sitting around on a hot summer afternoon trying to decide what to do, when Sue mentions that she has been thinking about an activity for next semester's Calculus classes. She explains that it involves a car trip during which odometer and speedometer readings are taken every thirty seconds. "Sounds like fun," says Bob, "Let's go." Car air conditioning on a hot summer afternoon aside, you are probably thinking, "these two desperately need to get a life." Sue grabs her laptop, watch and water bottle and heads for the car.

1. The trip odometer is set to zero, and timing begins. Although the road is not a straight line, we can imagine it as a thread, which we straighten to form single dimension travel.

    a) What does the odometer reading now represent?

    b) What does the speedometer reading represent?

2. At the start of the trip, the time is 0.0 minutes, and the odometer reading is 0.0 miles. At the end of the trip, the odometer reads 15.0 miles and 23.0 minutes have elapsed.

    a) Compute the average trip velocity $\bar{v} = \frac{\Delta s}{\Delta t}$ in miles per minute.

    b) If the car maintained a constant speed, and completed the trip in 23.0 minutes, what would the speedometer reading be?

3. On the next page is the graph of the position vs. time data points for the car trip.

   a) What would the slope of a secant line through any two of the data points represent?

   b) If you drew a smoothed curve through the data points, what physical quantity would the slope of a tangent line to the curve represent?

   c) How could you tell from the graph whether the car was moving when timing began?

   d) Was the car moving when the timing began?

   e) How could you tell from the graph when the car was moving the fastest?

   f) Approximate, from the graph, the time at which the speedometer reading is the highest.

   g) What characteristic of the graph would indicate that the car was slowing down?

   h) Approximate from the graph, over what time intervals the car is slowing down?

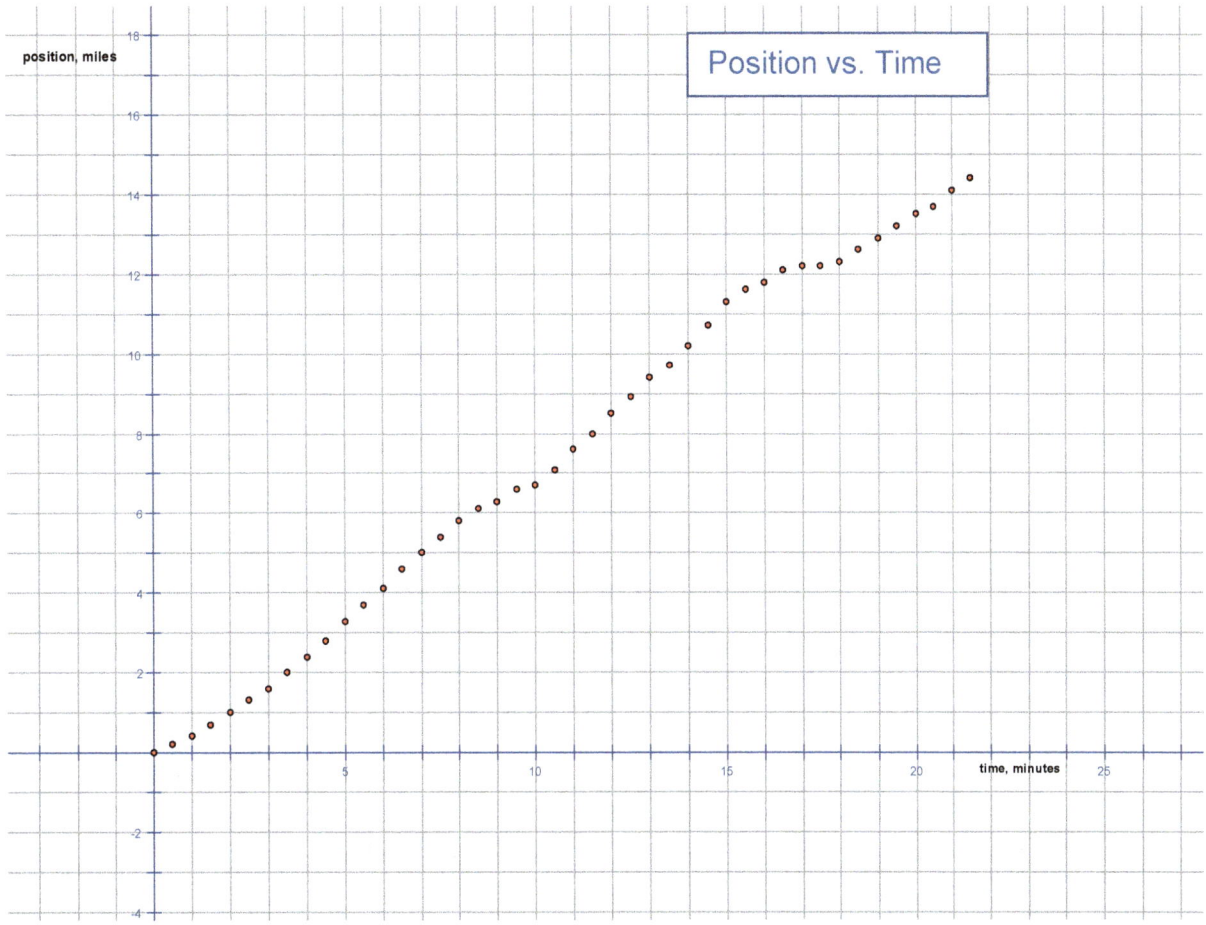

4. A single point of interest.

   Consider the position vs. time graph for the three points shown below.

   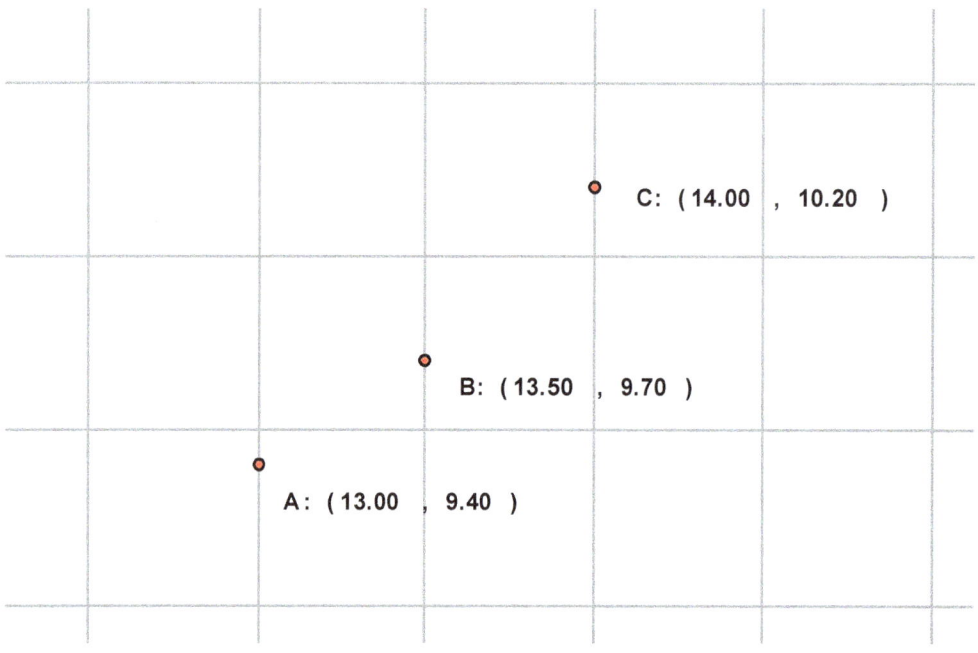

   a) Calculate the slope of the secant line through the points, A (13.0, 9.4) and B (13.5, 9.7).

   b) What does this quantity represent?

   c) Calculate the slope of the secant line through the points B (13.5, 9.7) and C (14, 10.2).

   d) What does this quantity represent?

   e) Calculate the average of these two slopes.

f) We will use the average of the two slopes to approximate the slope of the tangent at point B $(13.5, 9.7)$. Determine the equation of the line through the point $(13.5, 9.7)$ with this slope.

5. The table titled "Odometer Readings" contains the odometer readings for the trip. Using the procedure developed in problem 4, we compute the average velocity between each pair of data points. From this we approximate the instantaneous velocity at each time step.

    a) In the column "Change in s" calculate the difference in position between the two data points. For example, from $(0.5\ min, 0.2\ miles)$ to $(1.0\ min, 0.4\ miles)$, the change in position is $0.4\ miles - 0.2\ miles = 0.2\ miles$.

    b) In the column "Ave. v", divide the change in s by the change in t, (.5 min.). Continuing with our example the average velocity from 0.5 min to 1.0 min. is:

    $$\frac{Change\ in\ s}{Change\ in\ t} = \frac{0.2\ mile}{0.5\ minute} = \frac{0.4\ mile}{1\ minute}$$

    c) In the column "Approx. v," compute the average of the Ave. v for the given data point and the one following it. If we want the approximate velocity at the instant where $t = 0.5$ minutes, we average the two average velocities: 0.4 mi/min from 0.0 to 0.5 min, and 0.4 miles/min. from 0.5 to 1.0 min. The approximate velocity at $t = 0.5$ minutes is therefore 0.4 miles/min. Explain what this quantity represents.

Odometer Readings

Time (min.)	Position (miles)	Change In s (mi)	Ave. v (mi/min)	Approx. v (mi/min)	Time (min.)	Position (miles)	Change In s (mi)	Ave. v (mi/min)	Approx. v (mi/min)
0.0	0.0				11.5	8.0			
		0.2	0.4				0.5	1.0	
0.5	0.2			0.4	12.0	8.5			0.9
		0.2	0.4				0.4	0.8	
1.0	0.4			0.5	12.5	8.9			
		0.3	0.6				0.5	1.0	
1.5	0.7			0.6	13.0	9.4			
		0.3	0.6				0.3		
2.0	1.0			0.6	13.5	9.7			
		0.3	0.6				0.5		
2.5	1.3			0.6	14.0	10.2			
		0.3	0.6				0.5		
3.0	1.6			0.7	14.5	10.7			
		0.4	0.8				0.6		
3.5	2.0			0.8	15.0	11.3			
		0.4	0.8				0.3		
4.0	2.4			0.8	15.5	11.6			
		0.4	0.8						
4.5	2.8			0.9	16.0	11.8			
		0.5	1.0						
5.0	3.3			0.9	16.5	12.1			
		0.4	0.8						
5.5	3.7			0.8	17.0	12.2			
		0.4	0.8						
6.0	4.1			0.9	17.5	12.2			
		0.5	1.0						
6.5	4.6			0.9	18.0	12.3			
		0.4	0.8						
7.0	5.0			0.8	18.5	12.6			
		0.4	0.8						
7.5	5.4			0.8	19.0	12.9			
		0.4	0.8						
8.0	5.8			0.7	19.5	13.2			
		0.3	0.6						
8.5	6.1			0.5	20.0	13.5			
		0.2	0.4						
9.0	6.3			0.5	20.5	13.7			
		0.3	0.6						

9.5	6.6			0.4	**21.0**	14.1			
		0.1	0.2						
10.0	6.7			0.5	**21.5**	14.4			
		0.4	0.8						
10.5	7.1			0.9	**22.0**	14.6			
		0.5	1.0						
11.0	7.6			0.9	**22.5**	14.8			
		0.4	0.8						
11.5	8.0				**23.0**	15.0			

6. The graph on the next page presents the speedometer readings as a function of time for the trip, after they have been converted to miles per minute. On the same graph, are the approximate velocities computed from problem 5:

   a) Comment on how well the computed velocity matches the speedometer readings.

   b) Do you think that the method we used to compute the approximate velocities, that is also employed by your text, is a good way to approximate instantaneous rates of change for discrete data sets?

   c) On the graph of velocity vs. time, graph the line: $v = \bar{v}$ (average velocity) using the quantity you computed in problem 2.

   d) This line represents the average velocity over the whole trip. Is this line consistent with the speedometer readings, and/or the computed velocities?

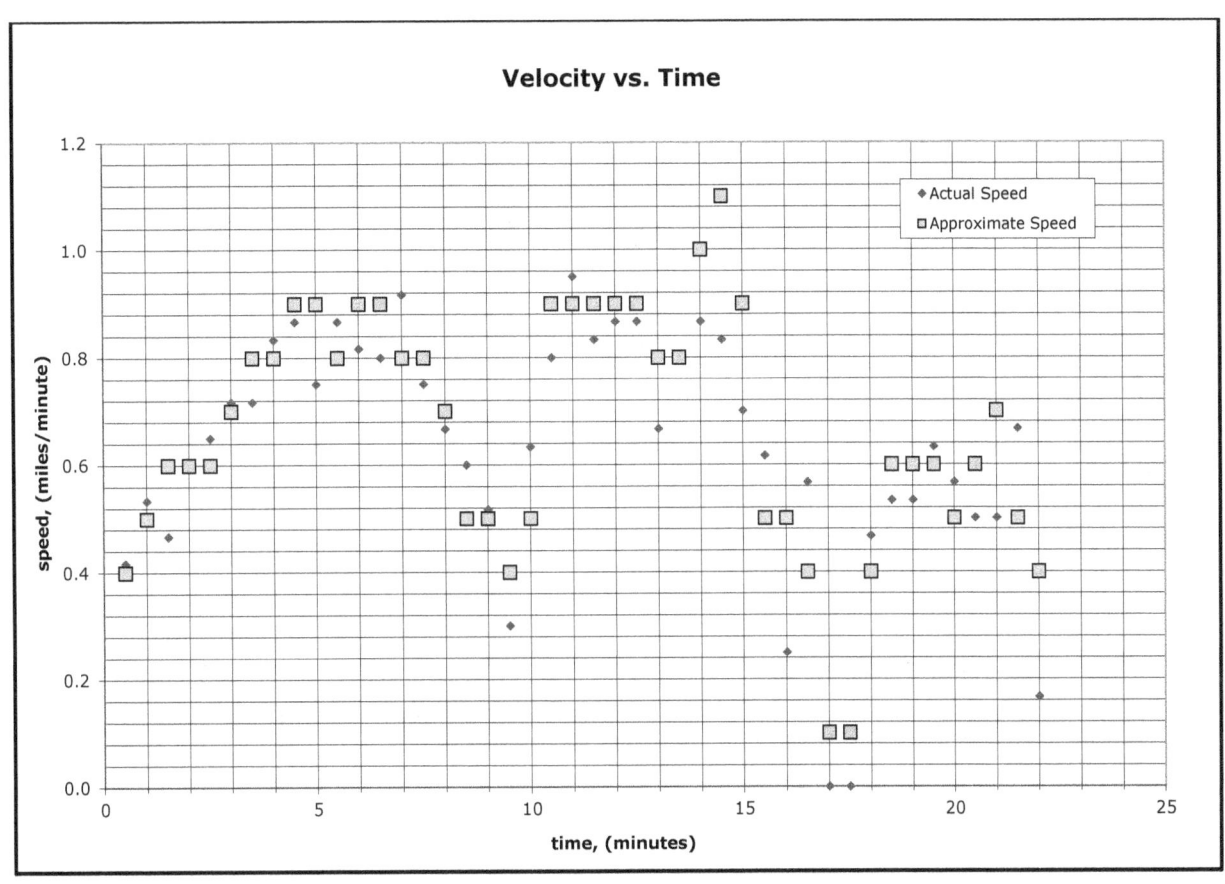

7. In problems 1 – 6 we explored the idea that given information about the position of a car at time, t, we can find the instantaneous rate at which that position is changing (velocity). We compared our calculations or approximate v's with the actual speedometer readings. The second big idea of Calculus is that given the rate of change of position (velocity), we can determine the net change in position by finding the area between the velocity vs. time graph and the t-axis.

   On the next page the graph has (i) the odometer readings in miles, and (ii) speedometer readings in miles per minute for the trip starting at time, $t = 0$, and ending with $t = 21.5$ minutes.

   a) How far has the car traveled in this time frame according to the odometer reading?

   b) Consider the velocity graph. The actual area of a 1mile/min × 1 min. square on the graph is $1.04\ cm^2$. The measured area under the velocity graph is $14.2\ cm^2$. This area represents the displacement of the car from 0 to 21.5 min. Compute this displacement in miles.

   Area EFGH = 1.04 cm²

   c) How does this compare with the ending odometer reading?

   d) If you were to graph the horizontal line representing the average velocity for the trip, and compute the area of the rectangle formed by this line and the t-axis between 0 and 21.5 minutes, what would you expect the area to be?

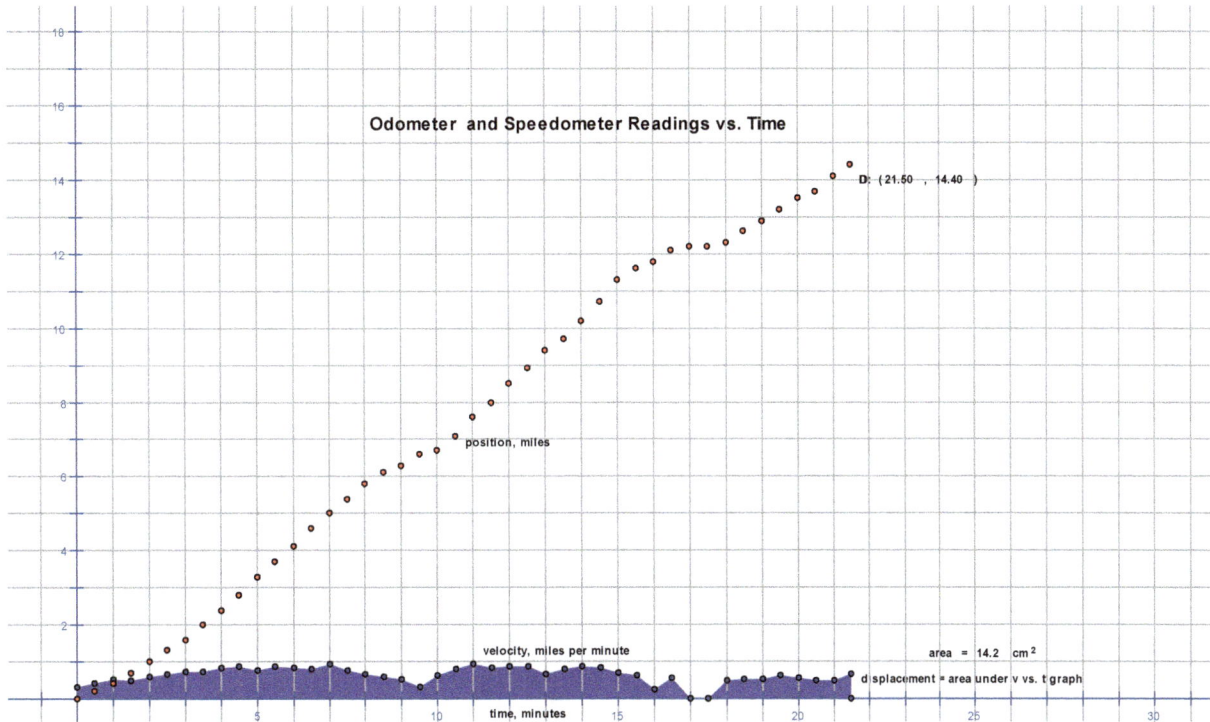

My Notes

Topic 7: Differentiation

Review Workshop: Differentiation Rules

The derivative of a function f at a point c, denoted $f'(c)$, is a number that gives us information about the function near c. In most applications of the derivative, we are interested in several values of the derivative at several distinct points. There are several efficient differentiation formulas to help us. We focus on them in this workshop.

Part 1.

1. Review the sum/difference and scalar multiple rules

   a) What is the differentiation rule for a sum or difference?

   b) What is the differentiation rule for a constant multiple?

   c) Use the appropriate formula to compute each derivative.

   1) $\frac{d}{dx}(x^2 + x^{1/4}) =$

   2) $\frac{d}{dx}(x^{-2} + 4x^3 - 10) =$

3) $\dfrac{d}{dx}\left(\dfrac{x+1}{\sqrt{x}}\right) =$

4) $\dfrac{d}{dx}(2\sin(x) - 4\cos(x)) =$

5) $\dfrac{d}{dx}\left(x^{0.3} - \sqrt{2x} + x^{-0.3}\right) =$

6) $\dfrac{d}{dx}\left(\sin\left(\dfrac{\pi}{4}\right) + 2^3\right) =$

7) If $f'(1) = 2$ and $g'(1) = -3$, calculate $(f - 2g)'(1)$.

2. Review the product rule.

   a) What is the product rule for differentiation? (First, write your answer using appropriate mathematical notation. Second, write your answer in words.)

   b) Use the appropriate formula(s) to compute each derivative.

   1) $\frac{d}{dx}(3x^2 + 4x - 1)(x^4 + x^3 + 13x^2 + 4x - 7) =$

   2) $\frac{d}{dx}\left(x + \frac{1}{\sqrt{x}}\right)\left(x - \frac{1}{x}\right) =$

   3) $\frac{d}{dx}(x^2 \cos(x)) =$

   4) $\frac{d}{dx}(\sin(x) \cos(x)) =$

   5) If $f'(0) = 1, g'(0) = -2, f(0) = 3$, and $g(0) = -1$, then $(f \cdot g)'(0) =$

6) If $f'(x), g'(x), h'(x)$ all exist, then the derivative of their product is
$(f \cdot g \cdot h)'(x)=$

3. Review the quotient rule.

   a) What is the quotient rule for differentiation? (First, write your answer using appropriate mathematical notation. Second, write your answer in words.)

   b) Use the appropriate formula(s) to compute each derivative.

   1) $\dfrac{d}{dx}\left(\dfrac{x^2-4}{x^4+16}\right) =$

   2) $\dfrac{d}{dx}\left(\dfrac{\sqrt{x}+x}{\sqrt{x}-x}\right) =$

   3) $\dfrac{d}{dx}\tan(x) = \dfrac{d}{dx}\left(\dfrac{\sin(x)}{\cos(x)}\right) =$

   4) $\dfrac{d}{dx}\csc(x) = \dfrac{d}{dx}\left(\dfrac{1}{\sin(x)}\right) =$

5) If $f'(2) = -2, g'(2) = 1, f(2) = 0$, and $g(2) = 3$, then $\left(\frac{f}{g}\right)'(2) =$

6) If $f'(x)$ exists and $f(6) = 0$, then $\frac{1}{f}(x) =$

Part 2.

Review all the formulas used in Part 1. Use the appropriate formula(s) to compute each derivative.

1. If $f(x) = 5x^4 - 10x^3 + 4x^2 + 8x + 4$, then $f'(x) =$

2. If $y = \frac{x - \cos(x)}{x + \sin(x)}$, then $y' =$

3. $\frac{d}{dx}(x \sin(x) \cos(x)) =$

4. If $f(x) = (x^2 + x^{-2})\left(\sqrt{x} + \frac{1}{\sqrt{x}}\right)$, then $f'(x) =$

5. If $f'(x), g'(x), and\ h'(x)$ exist, and $h(x) = 6$, then $\left(\frac{f \cdot g}{h}\right)'(x) =$

Part 3. Dispelling common differentiation errors.

1. Is the derivative of a product equal to the product of derivatives? Explain why or why not.

2. Is the derivative of a quotient equal to the quotient of the derivatives. Explain why or why not.

3. Add two more of your own questions about common differentiation errors to challenge your group.

Topic 8: The Chain Rule

Workshop: Composition of Functions and the Chain Rule

The chain rule of differentiation complements and, in certain cases, generalizes the basic rules of differentiation, such as the power rule or product rule. It provides a formula for differentiating the composition of two or more functions. Most functions used in calculus applications are compositions of two or more functions. Thus, the chain rule is an important and probably the most widely used rule of differentiation. Once understood, the chain rule gives a simple method for finding derivatives of very long and complicated expressions.

Part 1. Focus on terminology.

1. Define each term. Be sure to use appropriate mathematical notation in your answers.

    a) composition of two functions

    b) Chain Rule

Part 2A. Understanding the chain rule

Find each derivative. Recall that a function of the form $F(x) = \sin(x^2 + 4)$ can be considered as the composition $(g \cdot f)(x)$ of the following two functions: $f(x) = x^2 + 4$ and $g(x) = \sin(x)$.

3. $\frac{d}{dx} \sin(x^2 - 4) =$

2. $\frac{d}{dx}(1-x)^{-3} =$

3. $\dfrac{d}{dx}\tan^2(x) =$

4. $\dfrac{d}{dx}\sqrt{x^4 + 2x^2 + 4x} =$

5. $\dfrac{d}{dx}\cos(\cos(x)) =$

6. $\dfrac{d}{dx}(x^2 - x - 1)^{100} =$

Part 2B. Using the Chain Rule with multiple compositions.

1. Assuming that f, g and h are differentiable functions, find a chain rule formula for $F(x) = (h \cdot g \cdot f)(x)$.

2. $\dfrac{d}{dx}\sqrt{x + \sqrt{x + \sqrt{x}}} =$

3. $\frac{d}{dx}\sin^2(\sin^2(x)) =$

4. $\frac{d}{dx}[1 + (1 + x^2)^2]^{-2} =$

Part 2C. Deepening understanding of the chain rule.

Let the function T have derivative $T'(x) = (1 + x^2) - 1$ defined for every real number x.

1. Find the derivative of $T(1 + x^2)$.

2. Suppose $T(2) = 1$. Find the derivative of $T(T(1 + x^2))$ at $x = 1$.

Part 3. Using the appropriate differentiation rule.

Use any combination of differentiation rules to find each derivative.

1. Let $f(s) = x(3x - 4)^{2/3}$. Then $f'(x) =$

2. Suppose $f'(9) = -2$. Let $f(x) = f(4x + 1)$. Find $f'(2) =$

3. $\frac{d}{dx}\left(\frac{x+4}{x^2+1}\right)^4 =$

4. $\frac{d}{dx}\sqrt{x^4 + \cos^2(x)} =$

5. $\frac{d}{dx}(\tan^2(x) \cdot \sec^2(x)) =$

6. $\frac{d}{dx}\left(\frac{1}{4+x^2}\right)^8 =$

Part 4.

The problems in this section will help us discard any residual misconceptions concerning the Chain Rule.

Let $f(x) = x^2 - 1$ and $g(x) = x^2$. Find:

1. $h(x) = (g \cdot f)(x)$

2. $h'(x) = (g \cdot f)'(x)$

3. $f'(x)$

4. $g'(x)$

5. Circle the correct symbol: $h'(x) \quad = \quad \neq \quad g'(f'(x))$

6. Circle the correct symbol: $h'(x) \quad = \quad \neq \quad g'(x) \cdot f'(x)$

My Notes

Topic 9: Higher Order Derivatives and Implicit Differentiation

Review Workshop: A Comprehensive Look at Higher Order Derivatives and Implicit Differentiation

The derivative of a function, denoted by $f'(x)$, is itself a function, and as such may have its own derivative. If the derivative of the derivative of a function f exists, we call it the second derivative of f and denote it by $f''(x)$. The first and second derivatives of a function $y = f(x)$ are often used to obtain important information about the relationship between the dependent variable y and the independent variable x. They are key tools used in sketching the graph of $y = f(x)$. Most functions you are familiar with, such as polynomials, rational functions and trigonometric functions, can be differentiated several or even infinitely many times at most points. In a previous workshop, you used the first derivative to construct a linear approximation of a function near a point. If the first several derivatives of f are known at a point (i.e., $f(c)$, $f'(c)$, $f''(c)$, etc.), then it is possible to give a more accurate approximation of f near c using a polynomial. Your calculator uses this method to approximate the values of non-algebraic functions, such as trigonometric, exponential, or logarithmic functions.

The objective of this workshop is to enhance your differentiation techniques for computing higher order derivatives and derivatives of functions defined implicitly.

Part 1. Computing higher order derivatives

1. Let $f(x) = 2x^3 - 3x^2 + 4x + 10$. Find $f'(x), f''(x)$, and $f'''(x)$. What can you say about higher order derivatives such as $f^{(4)}(x), f^{(5)}(x)$, etc.? Explain.

2. Let $f(x) = \tan(x)$. Find $f'''(x)$.

3. Let $f(x) = a_0 + a_1 x + a_2 x^2 + a_3 x^3$. Show that $a_0 = f(0), a_1 = f'(0), a_2 = \frac{1}{2!}f''(0)$ and $a_3 = \frac{1}{3!}f'''(0)$.

4. The position of a particle moving on a line at time t is given by the formula $s(t) = t - \frac{1}{t^2}$. Find the velocity $v(t) = s'(t)$ and the acceleration $a(t) = s''(t)$ at the time when the position is zero (Hint: the position is zero when $s(t) = 0$, not necessarily when $t = 0$).

Part 2. Using patterns to find higher order derivatives.

1. Let $f(x) = \sin(x)$. Find $f'(x), f''(x), f'''(x)$, and $f^{(4)}(x)$.
   Use the pattern in your results to find $f^{(7)}(x), f^{(17)}(x)$, and $f^{(102)}(x)$.

2. Let $f(x) = \frac{1}{1+x} = (1+x)^{-1}$. Find $f'(x), f''(x), f'''(x)$, and $f^{(4)}(x)$.
   Use the pattern in your results to find $f^{(9)}(x)$, and $f^{(14)}(x)$.

3. Let f be a function satisfying $f'(x) = 2f(x)$. Find $f'(x)$, $f''(x)$, $f'''(x)$, and $f^{(4)}(x)$. Use the pattern to find $f^{(10)}(x)$ and a general formula for $f^{(n)}(x)$.

4. Let $f(x) = \frac{1}{x} = x^{-1}$. Find $f'(x)$, $f''(x)$, and $f'''(x)$. Use the pattern to calculate $f^{(99)}(x)$.

Part 3. Differentiating implicitly.

Suppose an equation defines y implicitly as a differentiable function of x. To differentiate y with respect to x implicitly (i.e., to find $y'(x)$) do the following:

   i. Differentiate both sides of the equation with respect to x, keeping in mind that y is a function of x. (Hint: it will be necessary to apply the chain rule.)
   ii. Solve the resulting (differential) equation for the unknown $y' = \frac{dy}{dx}$ in terms of x and y.

Solve the following problems:

1. Let $x^3 + y^3 = 3xy$ define y implicitly as a differentiable function of x. Find y'.

2. Let $\cos(x + y) = y$ define y implicitly as a differentiable function of x. Find y'.

3. Let $x = \sec(y)$ define y implicitly as a differentiable function of x. Find y'.

4. The volume V and the surface areas S of a sphere are related by the equation $36\pi V^2 = S^3$. Find $\frac{dS}{dV}$, the rate of change of the surface area with respect to the volume, when the volume is $\pi\sqrt{6}$ cubic feet.

5. For the previous problem, calculate $\frac{dV}{dS}$, the rate of change of the volume with respect to the surface area, when the volume is $\pi\sqrt{6}$ cubic feet.

6. Let $x^2 + y^2 = 1$ define y implicitly as a differentiable function of x. Find formulas for y' and y'' in terms of x and y only.

Part 4. Applying implicit differentiation.

Assume the equation $x = \sin(y)$ defines y implicitly as a differentiable function of x.

1. The point $\left(\frac{1}{2}, \frac{\pi}{6}\right)$ is on the graph of the equation $x = \sin(y)$. Verify this by showing that when $x = \frac{1}{2}$, $y = \frac{\pi}{6}$.

2. Find a formula for y' in terms of x and y, then find the slope of the tangent line at the point $\left(\frac{1}{2}, \frac{\pi}{6}\right)$.

3. Write the equation of the tangent line at the point $\left(\frac{1}{2}, \frac{\pi}{6}\right)$ in slope-intercept form $y = mx + b$. Next, replace y in the equation by $L(x)$ to obtain the linear function $L(x) = mx + b$.

4. Note that the graph of the tangent line $y = L(x)$ is very close to the graph of the equation $x = \sin(y)$ near the point $\left(\frac{1}{2}, \frac{\pi}{6}\right)$. Consequently, if x is close to 0.5, $L(x)$ will be close to y in the equation $x = \sin(y)$. Use the tangent line to approximate the y value in $x = \sin(y)$ when $x = 0.6$. This is called a *linear approximation*.

5. The numerical value found in the previous problem approximates $\arcsin(0.6)$. Use a calculator to compare the calculator value with the approximation found by using $L(0.6)$.

Topic 10: Related Rates

Review Workshop: Related Rates with Geometry and Modeling

Calculus provides tools for solving many practical problems. The objective of this workshop is to enhance your ability to solve problems involving related rates.

When solving a related rates problem we find the rate of change of one quantity by relating this quantity to other quantities whose values or rates of change are known. Some common formulas used in related rates problems are

- The Pythagorean Formula: $a^2 + b^2 = c^2$. Relates a and b.

- Area of circle: $A = \pi r^2$. Relates r and A.

- Volume of a sphere: $V = \left(\frac{4}{3}\right)\pi r^3$. Relates r and V.

- Volume of a cylinder: $V = \pi r^2 h$. Relates V and r or V and h.

- Area of a triangle: $A = \left(\frac{1}{2}\right)bh$. Relates A and b or A and h.

- Volume of a cone: $V = \left(\frac{1}{3}\right)\pi r^2 h$. Relates V and r or V and h.

Typically, to solve a related rates problem, one translates the verbal problem into a related mathematical problem, then uses calculus tools to solve the mathematical problem, and finally interprets the solution into the language of the original problem.

Part 1.

Review the rules of differentiation, including the chain rule, and answer the questions below.

1. Let $h(t)$ and $V(t)$ be differentiable functions of time t representing the height and volume of a box through the equation $V = h(20 - 2h)^2$.

   a) Use the chain rule to find a formula for $\frac{dV}{dt}$ in terms of h and $\frac{dh}{dt}$.

b) Find $\frac{dV}{dt}$ when $h = 2$ and $\frac{dh}{dt} = 1$.

2. Let V represent the volume of a cylinder, r the radius of the cylinder, and h the height of the cylinder.

   a) Determine an equation for V in terms of r and h.

   b) Suppose V, h and r are differentiable functions of time t. Use your formula from the previous problem and the chain rule to find a formula for $\frac{dV}{dt}$ in terms of r, h, $\frac{dr}{dt}$ and $\frac{dh}{dt}$.

   c) Determine V when $r = 2$ and $h = 3$

   d) Find $\frac{dr}{dt}$ when $\frac{dV}{dt} = 4, r = 2, \frac{dh}{dt} = -1$ and $h = 3$.

3. Let x and θ be differentiable functions of t and suppose $x = 200 \cot(\theta)$. Find a formula for $\frac{d\theta}{dt}$ in terms of $\frac{dx}{dt}$ and θ.

4. A rectangular box, with no top, is to be constructed from a 20-inch square piece of cardboard by cutting equal squares from each corner and then bending up the sides. We want to find a relation between the volume of the box and its height. Answer the questions below.

   a) Let l be the length of each side of the box and h the height of the box. Draw a picture of a square with a smaller square cut out of each corner. Label its dimensions in terms of l and h.

   b) Find a formula for volume V of the box in terms of l and h.

   c) Find an equation that relates the length l of one side of the square base of the box and the height h of the box.

   d) Use the equation found above to solve for the (extra) variable l in terms of h.

   e) Use the equation for l found above to substitute into the formula for the volume V of the box found previously to obtain a formula for the volume V in terms of the height h.

Part 2. A Modeling Approach

The general method for using mathematics to solve "real world" problems, including related rate problems, is summarized in the following diagram:

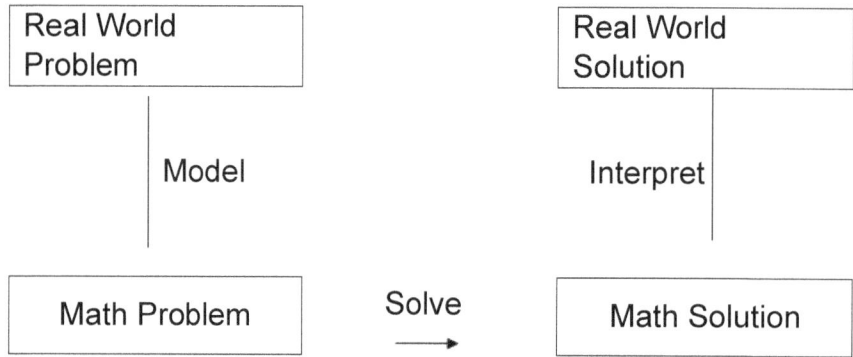

Suppose you are given a real-world problem, to which you desire a real world solution. To get there you must model the problem mathematically, solve the problem mathematically, and interpret your solution back into a real-world solution.

The following procedure can be used to solve related rates problems:

i. Carefully read the problem.
ii. Draw a schematic picture representing the problem, with labels if appropriate.
iii. Define the variables relevant to the statement of the problem.
iv. Find equation(s) that relate the variables. This can be done from the picture or from any known relationships among the variables. Common formulas are the Pythagorean Theorem or formulas for volume and area. The main equations are those which contain the variable whose rate of change is being sought. The other equations are auxiliary.
v. Determine the given (known) information. Generally, this is a rate of change.
vi. Determine what you want to find. This is also typically a rate of change.
vii. Use the auxiliary equation(s) to eliminate any extra variable by substituting for it into the main equations.
viii. Differentiate both sides of the main equation(s) with respect to time.
ix. Solve the resulting equation for the desired rate of change.

3. A conical tank, with vertex pointing down, is 20m across its circular top and 20m deep. Water is pumped into the tank at a rate of 40 cubic meters per minute. How fast is the level of the water rising when the water is 8m deep?

To solve this problem, complete the following steps:

a) Let h be the height of water in the tank at any time t, r the radius of the cone of water at depth h, and V the volume of water. Draw a picture (triangular view of cone) with labels r, h and the constants 20 meters across the top and height.

b) What rate(s) do we know (given information)?

c) What rate are we looking for?

d) Write an equation for the volume of water V in terms of r and h (main equation).

e) Write an auxiliary equation that relates h and r. (Hint: use similar triangles. A two-dimensional picture may be helpful.)

f) Use the auxiliary equation to solve for r in terms of h.

g) Substitute for r from the previous step into the main equation for V to obtain a relationship between V and h only.

h) Differentiate both sides with respect to time (t).

i) Solve for $\frac{dh}{dt}$.

j) Substitute values in for all the known rates and parameters to obtain a value for $\frac{dh}{dt}$.

k) How fast is the level of the water rising when the water is 8m deep?

2. A circular puddle is formed on the ground by water dripping in the center of the puddle. Suppose 5 square centimeters of water are added to the puddle each minute.

   a) Define variables for this problem. (Hint: use A and r)

   b) In words, what does $\frac{dr}{dt}$ represent?

   c) Discuss (no writing): Would you expect $\frac{dr}{dt}$ to be larger when the puddle is small or large?

d) Find a formula for $\frac{dr}{dt}$ in terms of $\frac{dA}{dt}$ and r.

e) Use your formula to calculate $\lim_{r \to \infty} \frac{dr}{dt}$. Write a single sentence interpretation of your answer. Does your result agree with your answer to the discussion question above?

f) Determine $\frac{dr}{dt}$ for the following values of r

  i. $r = 1$ cm

  ii. $r = 10$ cm

  iii. $r = 100$ cm

g) Do your answers to the previous problem agree with your answer to the discussion question?

3. A boat is being pulled to shore by a rope attached to a windlass atop a pier. The height of the windlass above the water is 6m, and the rope is being wound in at the rate of 5m per minute. How fast is the boat approaching the shore when it is 8m away?

4. A water trough has end pieces in the shape of inverted isosceles triangles with base 60cm and height 40cm. The trough is 400cm long. Water is being pumped into the trough at a rate of 9000 cubic centimeters per second. How fast is the level of the water in the trough rising when the water is 10cm deep?

Topic 11: Optimization

Review Workshop: Optimization with Geometry and Modeling

Calculus provides tools for solving many practical problems. The objective of this workshop is to enhance your ability to solve min-max problems. When solving a min-max problem we seek to obtain the "best" answer, which most often involves finding the maximum or minimum value of some function representing a quantity. For example, one may wish to determine the maximum profit or the minimum cost in a production line.

Typically, to solve a min-max problem, we translate a verbal problem into a related mathematical problem, then use calculus tools to solve the mathematical problem. Finally, the solution is interpreted back into the language of the original problem.

Part 1. Finding extreme values of a function

1. Define **extreme value** of a function f.

2. What does the Extreme Value Theorem state?

3. How can we find absolute extrema?

4. Let $f(x) = x - 3x^{1/3}$ be defined on the closed interval $I = [-1, 8]$. We want to find the absolute maximum and minimum of f on I. Answer each of the following questions to solve this problem.

77

a) The function f has an absolute max and absolute min on the interval $I = [-1, 8]$. Explain why?

b) Find $f''(x)$.

c) Find the only point x_1 in the *open* interval $(-1, 8)$ for which $f'(x_1) = 0$.

d) Find the (only) point x_2 in the *open* interval $(-1, 8)$ for which $f'(x_2)$ does not exist.

e) Compute the values $f(x_1), f(x_2), f(-1)$, and $f(8)$.

f) What is the absolute max and where does it occur?

g) What is the absolute min and where does it occur?

Part 2. A Modeling Approach

The general method for using mathematics to solve "real world" problems, including optimization problems, is summarized in the following diagram:

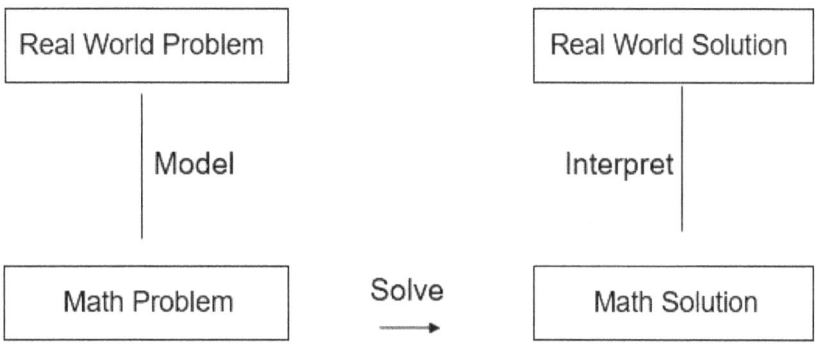

Suppose you are given a real-world problem, to which you desire a real-world solution. To get there you must model the problem mathematically, solve the problem mathematically, and interpret your solution back into a real-world solution.
The following procedure can be used to solve optimization problems:

i. Carefully read the problem.
ii. Draw a schematic picture representing the problem, with labels if appropriate.
iii. Define the variables relevant to the statement of the problem.
iv. Identify the variable for which the maximum/minimum is sought.
v. Find an equation for the max/min variable in terms of the other variables (main equation).
vi. Find any other equation(s) involving the independent variable and other extra variables (auxiliary equation(s)).
vii. Use the auxiliary equation(s) to substitute for the extra variable(s) into the main equation, thus obtaining a function that relates the max/min variable to a single independent variable.
viii. Determine the closed interval $[a, b]$ for the independent variable.
ix. Find the absolute max/min of the function.
x. Translate the solution into the language of the word problem. This may be as simple as giving a one sentence summary and adding units to the answer.

1. An open box is to be made from a square piece of cardboard measuring 20 in. on a side by cutting a square from each corner and folding up the sides. Find the dimensions for which the volume of the resulting box is a maximum.
   To solve the problem, follow the steps below:

   a) After reading the problem assign the letters V to the volume of the box, l to the length of the box, and x to the height of the box. Note that the width of the box is equal to the length (because it has a square base) and its height x is equal to the side of each cutout corner square.

   b) Draw a picture of the square with the corner squares removed and label each dimension.

   c) Volume V is the variable to be maximized. Find a formula for V in terms of l and x.

   d) Find an (auxiliary) equation that relates l to x.

   e) Solve for the extra variable l in the equation found in the previous step.

   f) Substitute for l in the formula for V found above and obtain a function V of one variable x.

g) Find the interval I of values for the independent variable x (x cannot be negative or greater than 10.) Why?

h) Find the absolute maximum of the function V for x ranging over the interval I found in the previous step.

i) The value of x which yields the maximum volume is the height of the box. What is the corresponding length l of the box?

2. A rectangular field is to have an area of 60,000 square meters. Fencing is required to enclose the field and divide it in half.

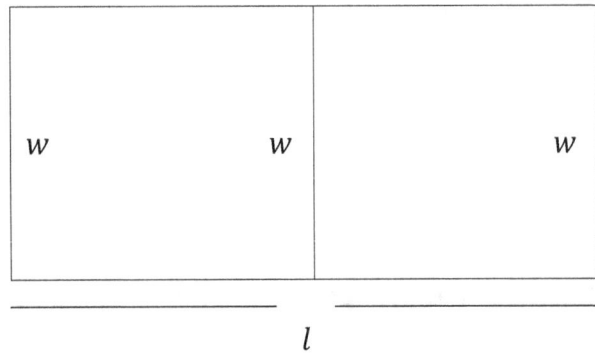

a) Express the total amount of fencing required as a function of the length of fence that divides the field (w in the diagram).

81

b) Find the minimum amount of fencing required.

c) What are the outer dimensions of the field that required the least fencing?

3. A closed rectangular container with a square base is to have a volume of 2250 cubic inches. The material for the top and bottom of the container will cost $2 per square in., and the material for the sides will cost $3 per square in. Find the dimensions of the container at the lowest cost.

4. A right circular cylinder is inscribed in a right circular cone so that the center lines of the cylinder and the cone coincide. The cone has a height of 6 and radius of base 3.

   a) Write down the formula for the volume of the cylinder in terms of its radius and height.

   b) Express the height of the cylinder as a function of its radius.

c) Express the volume of the cylinder as a function of its radius only.

d) Determine the radius of the cylinder that has maximum volume.

e) Determine the height and volume of the cylinder that has maximum volume.

My Notes

Topic 12: Surge Functions

Exploratory Workshop: Modeling with Surge Functions

This workshop will help you understand how calculus can be used in the analysis of Nicotine and Blood Alcohol concentrations.

Part 1.

Consider the graph of the rate of change of nicotine concentration in the blood over time, $C'(t)$, where t is measured in hours since smoking one cigarette and $C'(t)$ is measured in the milligrams per deciliter per hour.

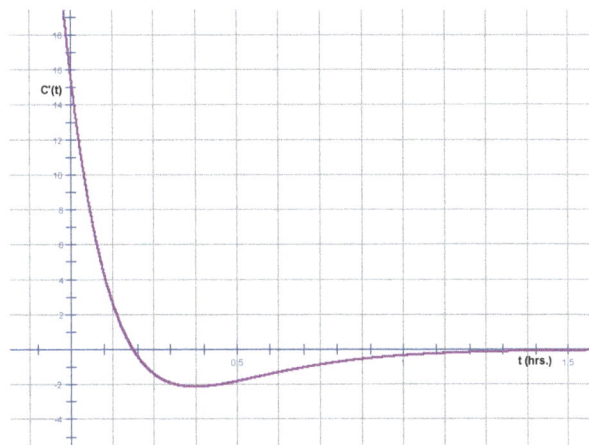

1. On the blank graph below, graph the concentration of nicotine in the blood as a function of time, assuming when $t = 0$ that $C = 0$ (at time 0, there is no nicotine in your blood).

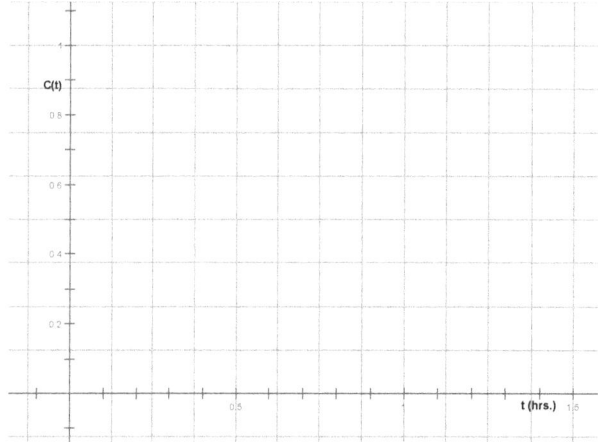

2. Give this some thought… is the concavity of your graph correct? How can you tell?

3. The correct graph for $C(t)$ is below. If your graph was not correct, where did you go wrong and why?

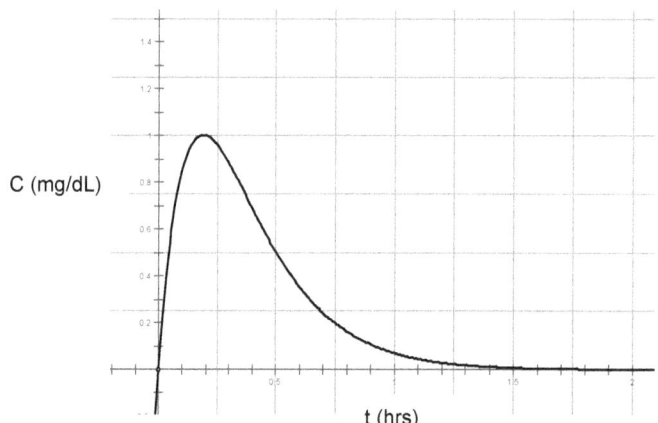

4. Based on the graphs of $C(t)$ and $C'(t)$, estimate the following:

   a) Find the time, t, where the maximum concentration occurs.

   b) Find the slope of the tangent line to the curve of $C(t)$ at $t = 0$.

5. Let's give some thought to the function for $C(t)$.

   a) What function would approximate $C(t)$ on the interval $[0, 0.1]$?

   b) What type of function would approximate $C(t)$ on the interval $[0.4, 2]$?

6. The function graphed above is $C(t) = 14.6te^{5.36t}$. Do the factors in this function correspond to the functions you described in 5a and 5b? Why or why not?

7. At what time is the nicotine leaving your body most rapidly? Answer:

   a) Using the graph of $C(t)$

   b) Using the graph of $C'(t)$

   c) Without using a graph (using calculus/algebra)

8. Explain in everyday English (to someone who is not in calculus) what the following mean in terms of the graphs:

   a) $C(t) > 0$

   b) $C'(t) > 0$

   c) $C''(t) > 0$

Part 2.

The functions of the form $y = Ate^{Bt}$ (like the above nicotine function $C(t)$) are called surge functions and display an initial approximate linear growth from zero that then (as t increases) become dominated by an exponential decay. Let's consider the general form of these functions, $y = Ate^{Bt}$.

1. On the graph below label the graphs of each of the following functions:

$$y_1 = 2te^{-0.2t} \qquad y_2 = 4te^{-0.2t} \qquad y_3 = 6te^{-0.2t}$$

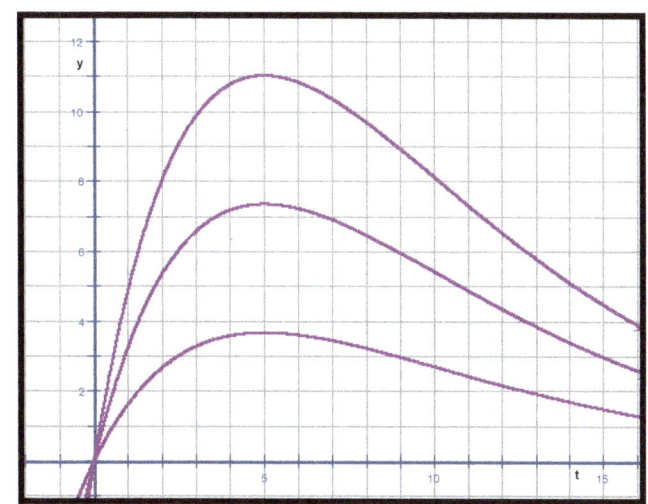

2. Briefly discuss what you notice about the shapes of these graphs and about what happens to the graphs as you change the parameter A in the equation. $y = Ate^{Bt}$.

3. For each of the three given functions, use calculus to find the slope of the tangent to the graph at $t = 0$.

4. For the general equation, $y = Ate^{Bt}$, what is the slope of the tangent to the curve at 0?

5. Verify your answer in #4 using calculus.

6. On the graph below label the graphs of each of the following functions:

$$y_1 = 2te^{-(1/5)t} \qquad y_2 = 2te^{-(1/8)t} \qquad y_3 = 2te^{-(1/10)t}$$

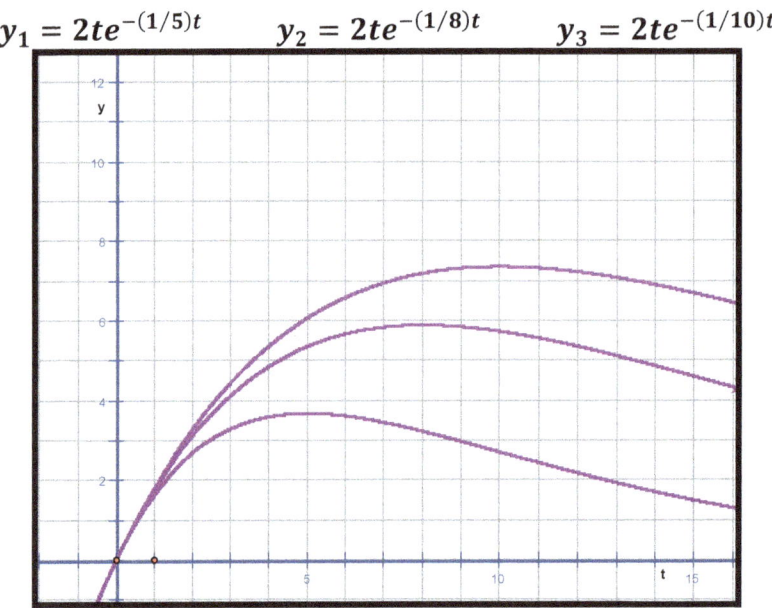

7. Briefly discuss what you notice about the shapes of these graphs and about what happens to the graphs as you change the parameter B in the equation $y = Ate^{Bt}$.

8. For which values of t does it appear that these functions take on their maximum value?

9. For each of the three given functions, use calculus to find the slope of the tangent to the graph at those values of t for which the function is a maximum.

90

10. Summarize your findings in the table:

Function	Value of t for y_{max}	Slope of tangent line at this point
$y = 2te^{-(1/5)t}$		
$y = 2te^{-(1/8)t}$		
$y = 2te^{-(1/10)t}$		
$y = Ate^{Bt}$		

11. Let's revisit the surge function for nicotine levels, $C(t) = 14.6te^{-5.36t}$. Based on what we now know,

   a) At what time, t, does the maximum concentration occur?

   b) What is the slope of the tangent line to the curve of $C(t)$ at $t = 0$?

   c) How do your answers compare with the answers you estimated from the graphs in Part 1, #4 (page 2)?

Part 3.

Now that we understand the A and B parameters in surge functions of the form $y = Ate^{Bt}$, let's examine blood alcohol level data and a possible surge function to model it.

The following data table shows the average of the values of Blood Alcohol Concentration (BAC) recorded over time for a group of male drinkers who rapidly consumed two drinks. For example, a value of 27 at a particular time means that the averages of the values of each participant's BAC at that time was .027%. The given data is plotted below, along with a possible mathematical model of the data in the form of a surge function.

Time (in minutes since the beginning of the drinking period)	0	10	20	30	40	50	60	90	120	150	180	210	240
Blood Alcohol Concentration (2 drinks)	0	27	32	42	43	38	34	26	20	12	9	6	2

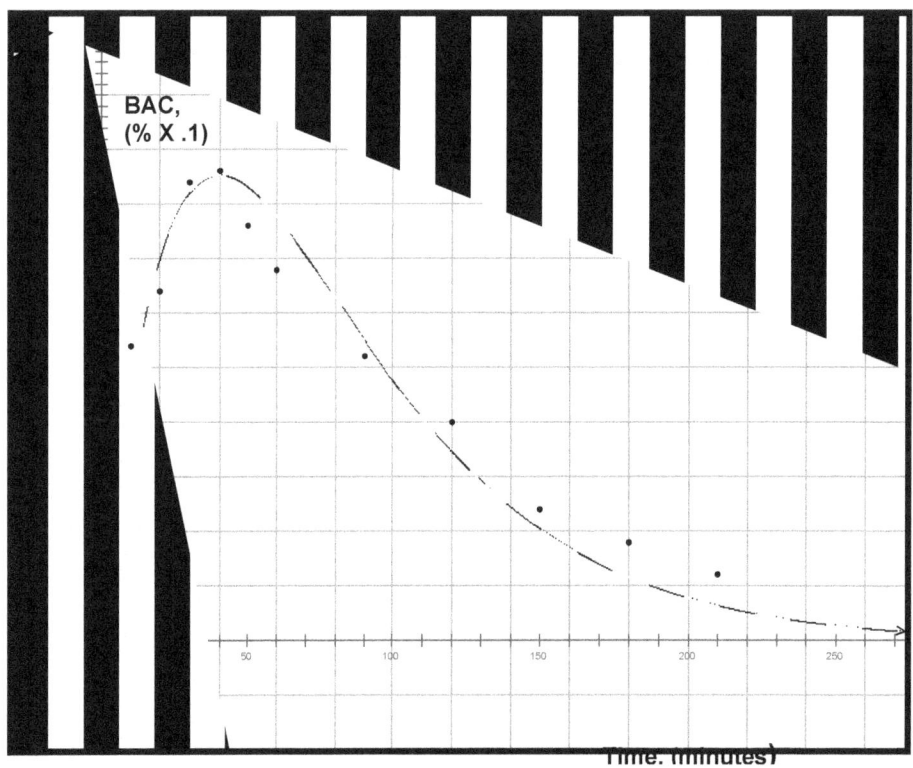

1. Using what you have learned about surge functions, write a possible equation that would model this situation.

2. Do you think that this is a good model for the given data? Why or why not?

Topic 13: Antiderivatives

Exploratory Workshop: Reversing Differentiation

Anti-differentiation, which is the process of finding anti-derivatives, is the "reverse" process of differentiation. An anti-derivative of a given function f is another function F, whose derivative is f: $F' = f$. Many practical problems are described by equations which contain one or more derivatives of a dependent variable. The solutions to these problems are obtained by anti-differentiating the dependent variable. Anti-differentiation is a central concept to the second half of calculus, namely, integral calculus.

We begin this workshop with a review of some important formulas and definitions. Then we review and apply the basic rules and properties of anti-differentiation.

1. Write the formula for each of the following:

   a) power rule for differentiation

   b) differentiation for the six basic trigonometric functions

   c) quotient rule

   d) product rule

   e) chain rule

2. Write your best definition for each term that follows.

   a) Antiderivative

   b) Indefinite integral

3. Verifying indefinite integrals

   Is there an error in the calculations (a-f) in the problems below? Check that the functions to the right of the equal sign are the correct indefinite integrals for the given integrands. Correct any errors you find.

   a) $\int \left(x^2 - \frac{1}{x^2}\right) dx = \frac{1}{3}x^3 + \frac{1}{x} + C$

   b) $\int \frac{1}{\sqrt{x+4}} dx = 2\sqrt{x+4} + C$

   c) $\int x\sqrt{x^2 + 4}\, dx = \frac{1}{3}(x^2 + 4)^{3/2} + C$

   d) $\int x \cos(x)\, dx = x \sin(x) + \cos(x) + C$

   e) $\int \sin^2(x)\, dx = \frac{1}{2}x - \frac{1}{2}\sin(x)\cos(x) + C$

   f) $\int \sin^2(x) \cos(x)\, dx = \frac{1}{3}\sin^3(x) + C$

4. We know that every differentiation formula can be written as an equivalent formula involving integrals. Complete the list below.

   a) $\int x^r dx = \frac{x^{r+1}}{r+1}$, for any number $r \neq -1$

   b) $\int \cos(x)dx =$

   c) $\int \sin(x)dx =$

   d) $\int \sec^2(x)dx =$

   e) $\int \csc^2(x)dx =$

   f) $\int \sec(x)\tan(x)dx =$

   g) $\int \csc(x)\cot(x)dx =$

5. To compute basic indefinite integrals we can use the two linearity properties of the definite integral:

   i. $\int (f(x) \pm g(x))dx = \int f(x)dx \pm \int g(x)dx$

   ii. $\int k \cdot f(x)dx = k\int f(x)dx$ for any constant k

   Use these two properties and known indefinite integrals to compute each of the following.

   a) $\int (x^2 + 2x - 4)dx =$

b) $\int \left(\sqrt{x} + \frac{1}{\sqrt{x}}\right) dx =$

c) $\int (3\sin(x) - 2\cos(x)) dx =$

d) $\int (\sec^2(x) + 1) dx =$

e) (e) $\int \left(\frac{x^4-1}{x^4}\right) dx =$

f) $\int (2x+3)(x^2-4) dx =$

g) $\int \frac{1+\sin(x)}{\cos^2(x)} dx =$

h) $\int \cot^2(x) dx =$

6. Explain how the chain rule and u-substitution can be used to show that
$$\int x \cos(x^2)\, dx = \frac{1}{2}\sin(x^2) + C$$

7. Compute the following anti-derivatives using u-substitution.

   a) $\int x^3(x^4 + 10)^7 dx =$

   b) $\int x^5 \cos(x^6 + 7) dx =$

   c) $\int (x+2)x^2 + 4x\, dx =$

   d) $\int \frac{\sin(\sqrt{x})}{\sqrt{x}} dx =$

e) $\int x\sqrt{x^2+1}\,dx =$

f) $\int x\sqrt{x+1}\,dx =$

g) $\int \sin(\sin(x))\cos(x)\,dx =$

h) $\int (2x+3)^{10}\,dx =$

8. Applying the anti-derivative

   a) Consider the following problem: An automobile manufacturer determines that the acceleration $a(t)$ under full throttle of one of its models is related to its velocity $v(t)$ by the differential equation $\frac{dv}{dt}(t) = 2/(3v(t)^{-1/2})$ (recall $a(t) = \frac{dv}{dt}(t)$). In this equation, v is in meters per second (m/s). Find the velocity $v(t)$ after t seconds if the velocity of the automobile at $t = 0$ is $v(0) = 9$m/s.

   Solve this problem by completing the following five steps:

   1) To find $v(t)$, write the differential equation as $v^{1/2}\frac{dv}{dt} = \frac{2}{3}$. Now write this in indefinite integral form: $\int v^{1/2} dv = \int \frac{2}{3} dt$.

   2) Compute each indefinite integral in the previous step (do not forget to add a C on one side).

   3) Use the initial condition $v(0) = 9$ to calculate the numerical value of C.
   ($v(0) = 9$ means when $t = 0, v = 9$)

   4) Substitute the value C into the equation found in 2.

   5) Solve for $v(t)$.

99

b) A population has size $P(t)$ at time $t > 0$ and grows according to the differential equation $\frac{dP}{dt}(t) = 4\sqrt{P(t)}$. Find $P(t)$ if $P(0) = 10$. Work with your group to devise the steps that are necessary, then solve the problem.

Topic 14: Indefinite Integrals

Review Workshop: Meeting Challenges of the Indefinite Integral

The function to be integrated is called the **integrand**. In $\int (3x + 1)dx$, the integrand is the function $f(x) = 3x + 1$.

If the integrand is a sum and/ or difference, use the addition rule $\int (f(x) \pm g(x))dx = \int f(x)dx \pm \int g(x)dx$ to create simpler integrals.

If the integrand is a product and/or quotient, you must either recognize it, (for example, $\int \sec(x)\tan(x)\,dx = \sec(x) + C$) or use a u-substitution.

Don't forget the constant "$+C$" when needed!

1. $\int (x^8 - 1)dx$

2. $\int (x - 1)^8 dx$

3. $\int x(x^2 + 1)^4 dx$

4. $\int \dfrac{1}{(4x-3)^8} dx$

5. $\int \sqrt{\frac{1}{2}x - 5}\, dx$

6. $\int \frac{x^2+6}{x^2}\, dx$

7. $\int \tan(x)\cos(x)\, dx$

8. $\int \frac{\tan(x)}{\cos(x)}\, dx$

9. $\int \frac{\sec(x)\tan(x)}{1+\tan^2(x)}\, dx$

10. $\int \frac{\sin(4x-1)}{1-\sin^2(4x-1)} dx$

11. $\int \frac{1+\cos(2x)}{\sin^2(2x)} dx$

12. $\int \frac{z+2}{\cos^2(z^2+4z-3)} dz$

13. $\int \frac{t^2 \cos(t^3-2)}{\sin^2(t^3-2)} dt$

14. $\int \frac{(6x-1)\sin(\sqrt{3x^2-x-1})}{\sqrt{3x^2-x-1}} dx$

My Notes

Topic 15: Riemann Sums

Review Workshop: Riemann Sums

In this workshop we review calculating Riemann Sums. We begin with a review of some important definitions, then we review calculating Riemann Sums in two multistep problems.

1. Define each term on the list below using the best definitions you have.

   a) Anti-derivative

   b) Indefinite integral

   c) Power rule for the anti-derivative

   d) *U*-substitution formula for anti-differentiation

   e) Riemann Sum

2. How can we approximate area by using Riemann Sums?

3. Follow the steps to calculate the Riemann Sum.

   a) Graph the function $y = x^2 + 2x + 1$, with attention to the interval $0 \leq x \leq 1$.

   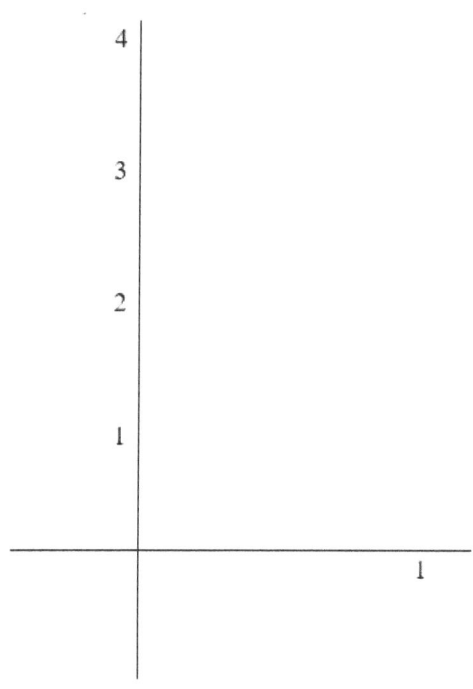

   b) Divide the interval $0 \leq x \leq 1$ into two parts.
      Draw two rectangles on these two intervals whose areas are "too big" for the area under the curve.
      Find the areas of these rectangles. (What is the height of each rectangle?) Their sum is an <u>upper bound</u> for the area under the curve.

c) Now draw two rectangles on these two intervals whose areas are "too small" for the area under the curve.
   Find the areas of these rectangles. Their sum is a lower bound for the area under the curve.

d) For the same function as above, divide the interval $0 \leq x \leq 1$ into four parts. Draw four rectangles on these four intervals whose areas are "too big" for the area under the curve.
   Find the areas of these rectangles. Their sum is another upper bound for the area under the curve. Does this seem to be a better approximation than the two-rectangle upper bound?

e) Now draw four rectangles on these four intervals whose areas are "too small" for the area under the curve.
   Find the areas of these rectangles. Their sum is another lower bound for the area under the curve. Does this seem to be a better approximation than the two-rectangle lower bound?

f) Repeat the process for six "too big" rectangles and six "too small" rectangles. What are the upper bound and the lower bound for the area under the curve?

g) Repeat the process for *n* "too small" rectangles.

h) What is the first rectangle's width? What is the first rectangle's height? What is the first rectangle's area?

i) What is the second rectangle's width? What is the second rectangle's height? Leave the area as powers of $\frac{1}{n}$. (For example, $a\left(\frac{1}{n}\right)^3 + b\left(\frac{1}{n}\right)^2 + \cdots$ and so on.)

j) is the third rectangle's width? What is the third rectangle's height? Leave the area as powers of $\frac{1}{n}$. (For example, $a\left(\frac{1}{n}\right)^3 + b\left(\frac{1}{n}\right)^2 + \cdots$ and so on.)

k) What is the *n*-th rectangle's width? What is the *n*-th rectangle's height?

l) Now add the *n* areas. Gather powers of $\frac{1}{n}$.

m) Simplify the area of *n* rectangles using the formulas:

i) $1 + 2 + 3 + 4 + \cdots + n = \frac{n(n+1)}{2}$

ii) $1^2 + 2^2 + 3^2 + 4^2 + \cdots + n^2 = \frac{n(n+1)(2n+1)}{6}$

iii) $1^3 + 2^3 + 3^3 + 4^3 + \cdots + n^3 = \left[\frac{n(n+1)}{2}\right]^2$

n) Find the limit of the area as $n \to \infty$.

4. Follow the steps to calculate the Riemann Sum.

   g) Let $f(x) = \sqrt{1-x^2}$ for x in the interval $[0,1]$. We want to approximate the area under the graph of $y = f(x)$ bounded by the x and y axes.
   Sketch the graph of $y = f(x)$ on the interval $[0,1]$. (This is a quarter of a circle centered at the origin with radius 1.)

   h) Divide the interval $[0,1]$ into $n = 2$ subintervals $\left[0, \frac{1}{2}\right]$ and $\left[\frac{1}{2}, 1\right]$. Compute the lower Riemann Sum
   $$L_2(f) = f(1/2) \cdot (1/2) + f(1) \cdot (1/2).$$

   Also, draw a picture showing the area represented by L_2.

i) Divide the interval [0,1] into $n = 2$ subintervals $\left[0, \frac{1}{2}\right]$ and $\left[\frac{1}{2}, 1\right]$. Compute the upper Riemann Sum
$$U_2(f) = f(0) \cdot (1/2) + f(1/2) \cdot (1/2).$$

Also, draw a picture showing the area represented by U_2.

j) Now divide the interval [0,1] into $n = 4$ subintervals: $\left[0, \frac{1}{4}\right], \left[\frac{1}{4}, \frac{1}{2}\right], \left[\frac{1}{2}, \frac{3}{4}\right], \left[\frac{3}{4}, 1\right]$.
Compute the lower Riemann Sum
$$L_4(f) = f(1/4) \cdot (1/4) + f(1/2) \cdot (1/4) + f(3/4) \cdot (1/4) + f(1) \cdot (1/4).$$

Also, draw a picture showing the area represented by L_4.

k) Now divide the interval $[0,1]$ into $n = 4$ subintervals: $\left[0, \frac{1}{4}\right], \left[\frac{1}{4}, \frac{1}{2}\right], \left[\frac{1}{2}, \frac{3}{4}\right], \left[\frac{3}{4}, 1\right]$.

Compute the upper Riemann Sum
$$U_4(f) = f(0) \cdot (1/4) + f(1/4) \cdot (1/4) + f(1/2) \cdot (1/4) + f(3/4) \cdot (1/4)$$

Also, draw a picture showing the area represented by U_4.

l) Repeat the process above for $n = 8$ to find $L_8(f)$ and $U_8(f)$.

m) Compare the values of $L_2(f)$, $L_4(f)$, and $L_8(f)$ together.

Compare the values of $U_2(f)$, $U_4(f)$, and $U_8(f)$ together.

n) Determine the exact area under the curve using the formula for the area of a circle.

o) How does the exact value of the area compare with the approximate values $L_2(f)$, $L_4(f)$, $L_8(f)$ and $U_2(f)$, $U_4(f)$, $U_8(f)$?

p) What can you conclude about the approximations U_n and L_n as the number n of divisions of $[0,1]$ increases?

My Notes

Topic 16: The Fundamental Theorem of Calculus

Review Workshop: The FTC

The Fundamental Theorem of Calculus provides an easy way for computing the definite integral for many functions. In this workshop we review calculating the definite integral using versions of the Fundamental Theorem of Calculus.

1.
   a) What is the first version of the Fundamental Theorem of Calculus?

   b) Complete the five properties of the definite integral:

   i. $\int_a^b (f(x) \pm g(x))dx =$

   ii. $\int_a^b (k \cdot f(x))dx =$

   iii. $\int_a^b (f(x))dx =$

   iv. $\int_a^a (f(x))dx =$

   v. $\int_b^b (f(x))dx =$

2. Compute each definite integral:

   a) $\int_{-1}^{2}(x^3 - x)dx =$

   b) $\int_{0}^{4}(x^2 - 6x + 3)dx =$

   c) $\int_{4}^{9}\left(\sqrt{x} - \frac{1}{\sqrt{x}}\right)dx =$

   d) $\int_{0}^{\pi}(\sin(x))dx =$

   e) $\int_{0}^{\pi/4}(\sec(x)\tan(x))dx =$

   f) $\int_{5}^{7}(f(x))dx$ given $\int_{5}^{7}(f(x))dx = 3$

g) $\int_1^5 (f(x))dx$ given $\int_1^7 (f(x))dx = 5$ and $\int_5^7 (f(x))dx = 3$.

h) $\int_5^7 (2f(x) - 4g(x))dx$ given $\int_5^7 (f(x))dx = 3$ and $\int_5^7 (g(x))dx = 4$.

3. How can we compute the definite integral by *u*-substitution?

4. Compute each definite integral:

   a) $\int_0^2 (\sqrt{4x+1})dx =$

   b) $\int_0^{\sqrt{x}} (x\sin(x^2))dx =$

   c) $\int_1^2 \left(\dfrac{4x}{(2x^2-1)^3}\right)dx =$

d) $\int_0^{\pi/4} \left(\sin(x)\sqrt{\cos(x)}\right) dx =$

e) $\int_0^1 \left(\frac{1}{(4-x)^2}\right) dx =$

f) $\int_0^{\pi/4} (\tan(x)\sec^2(x)) dx =$

5. How can we use the definite integral to calculate the area of a plane region?

6. Compute the area of the plane region between the graph of

(a) $f(x) = x\sqrt{x^2 + 1}$, the x-axis, $x = 0$ and $x = 1$

(b) $f(x) = 1/x^2$, $g(x) = x^{2/3}$ and the vertical lines $x = 1$ and $x = 8$

(c) $y = x^3$ and $y = x$

(d) $x = 4 - y^2$ and $x = y - 2$.

7. How can we use the second version of the Fundamental Theorem of Calculus to compute the derivative of a function?

8. Calculate each of the following derivatives.

a) $\frac{d}{dx} \int_0^x (\sqrt{t^4 - 1}) dt =$

b) $\frac{d}{dx} \int_2^x \left(\frac{1}{1+t^3}\right) dt =$

c) $\frac{d}{dx}\int_x^1 (1-t^4)dt =$

d) $\frac{d}{dx}\int_1^{1/x} \left(\frac{1}{t}\right) dt =$

e) $\frac{d}{dx}\int_{2x}^{4x} (\sin^2 t)dt =$

f) $\frac{d}{dx}\int_{\sin(x)}^{\cos(x)} \left(\frac{1}{1+t^2}\right) dt =$

My Notes

My Notes

www.ingramcontent.com/pod-product-compliance
Lightning Source LLC
Chambersburg PA
CBHW081430070526
44586CB00020B/2541